Selected Works by JD Bernal
Prism Key Press | www.prismkeypress.com

ISBN: 978-1463693411

Selected Works
JD Bernal

Contents

The World, The Flesh and The Devil

I. The Future

There are two futures, the future of desire and the future of fate, and man's reason has never learnt to separate them. Desire, the strongest thing in the world, is itself all future, and it is not for nothing that in all the religions the motive is always forwards to an endless futurity of bliss or annihilation. Now that religion gives place to science the paradiscial future of the soul fades before the Utopian future of the species, and still the future rules. But always there is, on the other side, destiny, that which inevitably will happen, a future here concerned not as the other was with man and his desires, but blindly and inexorably with the whole universe of space and time. The Buddhist seeks to escape from the Wheel of Life and Death, the Christian passes through them in the faith of another world to come, the modern reformer, as unrealistic but less imaginative, demands his chosen future in this world of men.

Can we in any better way reconcile desire and fate? In the belief of the scientist the future can yield to an objective analysis only if he can put aside all desire of one future or another; and yet, in reaching for this unattainable understanding, by some mutual influence his desires and the events may grow more and more into harmony. Holding this hope, or better still, moved by a pure curiosity for things to come, how is it possible to examine scientifically the future? For in the science of the future observation is as impossible as experiment; and of the three methods there is left to us only prediction. In the other sciences prediction plays but a small part, and rightly so, for verification follows closely on its heels; but there are general methods in scientific prediction and we may try to apply them in dealing with the whole future state.

First and always, it is necessary to exclude as far as

possible, illusion; for to most of us the future is the compensation and fulfilment of all that the present and the past have lacked; and the future being unknown and incontrovertible has been a fair ground on which to place all these hopes and desires. But in scientific prediction these desires are the most delusive guides. The opposite danger is as great and more insidious: in our lives we take the present for granted to an extend far greater than we can realize, so that even when we are thinking of the future we cannot separate the historic accidents of the society in which we were born from the axiomatic bases of the universe. Until the last few centuries this inability to see the future except as a continuation of the present prevented any but mystical anticipations of it. Luckily these complementary errors affect different parts of the future. It is in the near future where we are still sympathetically related to men and events that our desires have the most power to twist our appreciation of facts. We care less about the more distant future, but to approach it at all we must divest ourselves of so many customary forms, that even the more enlightened prophets lets their imagination stop in some static Utopia in despite of all evidence pointing to ever increasing acceleration of change.

What positive ideas can be found to take the place of the naïve anticipation that the future will be like the present but more pleasant (or more unpleasant according to one's disposition)? The leading principle is that by which Lyell founded scientific geology: the state of the present and the forces operating in it contain implicitly the future state and point the way to its interpretation. We have three disciplines of thought to help us to this interpretation. History (of which human history is only a minimal part) tells us how things have changed and how by inference they will change in the future. Strictly, prophecy should be treated as part of history, but, until history has found its laws, it must chiefly be used as a storehouse of illustrative facts; though one might say loosely that everything that will happen must conform with the spirit of History. The physical sciences, as far as we know them, give us the material of which future as much as past is built, and the manner of that building. The manner appears to us as physical law but it may well be found to be a tautology which we are congenitally too limited to grasp. Lastly there is the knowledge of our desires, but though the future according to our desires, is an illusion, our desire are, paradoxically, already tending to be the chief agent of change in the universe; it is only that the actual change is so rarely the desired change.

The initial difficulty in the general prediction of the future is its enormous complexity and the interdependence of all its parts; but this complexity is not completely chaotic and we can always attack it by considering it as a product of chance and determinism, chance where we cannot see relationships, determinism where we can. The events out of which so complicated a thing as the general state of the universe is built, form neither one indivisible whole nor a set of equally independent units, but consist of complexes (nebula, planet, sea, animal, society) of which the components are themselves complex parts. This hierarchy of complexes is not imagined to have any objective validity, it is only an expression of the modes of human thought, a convenient simplification which makes science possible. Inside each complex, development proceeds according to its own rules, determined by the nature of the complex; but these rules always include, if they do not entirely reduce to, what is, in effect, the statistical chance interaction of complexes of a lower order. The death-rate of a town, for instance, can be shown to be a function of the amount of money it spends on sanitary measures, but the individual deaths appear, from the point of view of the town, to be due to chance circumstances, though again for each individual concerned they are determined. We can always leave out the higher complexes when we are considering the lower. An atom of oxygen will respond to its environment in the same way in a nebula, in a rock, or in a human brain.

Now the complex we are concerned with here is the human mind, and so we can fairly start with the assumption that the rest of the universe goes on its way determined by its physical, chemical and biological laws except in so far as man himself intervenes. Absolutely, we know hardly anything of these laws, but relatively to our knowledge of human behavior we know them so well that the future they present - the astronomical, geological, biological future - seems a fixed and stable thing.

In human affairs the immediate future reveals itself in the following of tendencies visible in the present; beyond that must come the application and development of present knowledge. This is the minimal basis for prediction; but our present knowledge carries with it the implication of still further advances in knowledge along the same lines. It is the applications of this new knowledge and the secondary results that flow from them that will chiefly concern us, because it is clearly impossible to go further and include unimagined discovery. Of course, there is a considerable chance that one of the unpredictable discoveries will be so important that it will turn aside the whole course of development. But to be deterred by this chance would be to abandon any attempt at prediction. Already the chance element comes in when we consider applications or developments of knowledge in more than one restricted field; because although we can predict the development in that field fairly well, we cannot predict the rate of development; and so the rates of development in different fields, which are constantly reacting on each other, being unpredictable, the resultant future becomes more and more uncertain the father we look forward. The only way to deal with this complexity is by separating the variables as best we can, by arbitrarily considering developments as proceeding in one field without any developments in any of the others, and then combining the results attained by applying this method in different fields. At the same time we must keep in mind that the state of development at any one time period must be a self-consistent whole. Each line of development must have reached the level which is implied by the necessities of any of the other lines: for instance, the chemical control of life requires the development of chemical technique and apparatus of a very high order. On the other hand, whole sections of certain developments may become superfluous owing to developments in other fields; for instance, the manufacture of synthetic food and the industry connected with it would be unnecessary if blood were used directly as the motive power for animals.

Obviously we cannot proceed with this method in detail: if we could, we should not only be able to predict the future exactly, but to make it the present. For brevity, it is worth considering three fields only.

Man is occupied and has been persistently occupied since his separate evolution, with three kinds of struggle: first with the massive, unintelligent forces of nature, heat and cold, winds, rivers, matter and energy; secondly, with the things closer to him, animals and plants, his own body, its health and disease; and lastly, with his desires and fears, his imaginations and stupidities. In each of these divisions in turn we will make the arbitrary assumption that his progress in it will continue while in other respects he remains the same.

II. The World

First, then, in the material world. Here prediction is on its surest ground, and is, in the first stages, almost a business of mathematics. The physical discoveries of the last twenty-five years must find their application in the world of action - a process which has hardly begun, but the nature of which can easily be seen. So far we have been living on the discoveries of the early and mid-nineteenth century, a macro-mechanical age of power and metal. Essentially it succeeded in substituting mechanism for some of the simpler mechanical movements of the human body, with steam and later electrical power in the place of muscle energy. This was sufficient to revolutionize the whole of human life and to turn the balance definitely for man against the gross natural forces; but the discoveries of the twentieth century, particularly the micro-mechanics of the Quantum Theory which touch on the nature of matter itself, are far more fundamental and must in time produce far more important results. The first step will be the development of new materials and new processes in which physics, chemistry and mechanics will be inextricably fused. The stage should soon be reached when materials can be produced which are not merely modifications of what nature has given us in the way of stones, metals, woods and fibers, but are made to specifications of a molecular architecture. Already we know all the varieties of atoms; we are beginning to know the forces that bind them together; soon we shall be doing this in a way to suit our own purposes. In fact, Professor Goldschmidt of Oslo has already made many model structures in which existing substances are closely copied in different atoms, so as to make new substances, softer or harder, or more or less fusible. Sulpho-nitrdes with silicate structures will be harder and more infusible than anything on earth. A similar substance - carboloy - which is already on the market - combines the strength of steel with the hardness of diamond, and is capable of working glass like a metal. There are similar possible model structures for organic substances; the complexities are greater but the results will be more far-reaching. The linked molecules that make fibers and elastic substances such a rubber or muscle, are already yielding to X-ray investigation; the proteid bodies of living matter must have an analogous but more complex structure. After the analysis will come the synthesis; and for one place in which we can imitate nature we will be able to improve on her in ten, and furnish models of organic materials with more varied properties and capable of withstanding more rigorous conditions. The result - not so very distant - will probably be the passing of the age of metals and all that it implies - mines, furnaces, and engines of massive construction. Instead we should have a world of fabric materials, light and elastic, strong only for the purposes for which they are being used, a world which will imitate the balanced perfection of a living body.

At the same time, much that we require for the purposes of modern life would become no longer necessary. With improved systems of chemical manufacture our food and our clothing will be made with much less expenditure of energy in manufacture and transport. And the development of mechanism will not cease: it should turn into more refined forms - heat-engines capable of working at lower and lower temperature differences, engines of higher and higher speed, electrical machines of high potential and high frequency - and should lead to the solution of two most fundamental problems, the efficient transmission of energy by low frequency (wireless) waves, and the direct utilization of the high frequency (light) waves of the sun. On the chemical side the problem of the production of food under controlled conditions, biochemical and ultimately chemical, should become an accomplished fact. In the new synthetic foods will be combined physiological efficacy and a range of flavor equal to that which nature provides, and exceeding it as taste demands; with a range of texture also, the lack of which so far has been the chief disadvantage of substitute food stuffs. With such a variety of combinations to work on, gastronomy will, for the first time, be able to rank with the other arts.

All these developments would lead to a world incomparably more efficient and richer than the present, capable of supporting a much larger population, secure from want and having ample leisure, but still a world limited in space to the surface of the globe and in time to the caprices of geological epochs. Already ambition is stirring in men to conquer space as they conquered the air, and this ambition - at first fantastic - as time goes on become more and more reinforced by necessity. Ultimately it would seem impossible that it should not be solved. Our opponent is here the simple curvature of space-time - a mere matter of acquiring sufficient acceleration on our own part -

which, sooner or later, must be practicable. Even now it is possible to imagine methods of accomplishing it, based on no more knowledge than we already possess. The problem of the conquest of space is one in which all the difficulties are at the beginning. Once the earth's gravitational field is overcome, development must follow with immense rapidity. Without going too closely into the mechanical details, it appears that the most effective method is based on the principle of the rocket, and the difficulty, as it exists, is simply that of projecting the particles, whose recoil is being utilized, with the greatest possible velocity, so at to economize both energy and the amount of matter required for propulsion. Up to the present all forms of rocket depend on the movement of masses of gas in which the individual molecules are moving at high velocities in perfectly random directions, and use is only made of the average velocity in the desired direction. What is wanted in the first place is a form of Maxwell's Demon which will allow only those molecules, whose velocities are high and in the direction opposite to the trajectory of the rocket, to escape. The next difficulty is that to set in motion any large rocket the mass of gas required is of the same order as the weight of the rocket itself, so that it is difficult to imagine how the rocket could contain enough material to maintain its propulsion for any length of time. When the radio-transmission of energy is effected half the difficulty will be removed and the projection may very well ultimately be effected by means of positive rays at high potential. It may be that both the problem of space travel and the ethereal transference of energy have already been solved by Professor Japolsky's magnetofugal waves. These are a type of magnetic vortex ring, propagated through space, which, instead of spreading as ordinary electromagnetic waves, remain concentrated along the axis of propagation. Apart from its mode of projection, the construction of the space vessel offers little difficulty since it is essentially the same problem as that of the submarine. Naturally the first space vessels will be extremely cramped and uncomfortable, but they will be manned only by enthusiasts. The problem of landing on any other plant or of returning to earth is much more difficult, mainly because it requires such a nice control of acceleration. Probably the first journeys will be purely for exploration, without landing, and the travellers, if they return to earth at all, will have to abandon their machine and descend in parachutes.

However it is effected, the first leaving of the earth will have provided us with the means of travelling through space with considerable acceleration and, therefore, the possibility of obtaining great velocities - even if the acceleration can only be maintained for a short time. If the problem of the utilization of solar energy has by that time been solved, the movement of these space vessels can be maintained indefinitely. Failing this, a form of space sailing might be developed which used the repulsive effect of the sun's rays instead of wind. A space vessel spreading its large, metallic wings, acres in extent, to the full, might be blown to the limit of Neptune's orbit. Then, to increase its speed, it would tack, close-hauled, down the gravitational field, spreading full sail again as it rushed past the sun.

So far, those who have considered spatial navigation have regarded it from the point of view of exploration and planetary visitation, but the vast importance of escaping from the earth's gravitational field has been almost entirely overlooked. On earth, even if we should use all the solar energy which we received, we should still be wasting all but one two-billionths of the energy that the sun gives out. Consequently, when we have learnt to live on this solar energy and also to emancipate ourselves from the earth's surface, the possibilities of the spread of humanity will be multiplied accordingly. We can imagine this occurring in definite stages. When the technicalities of space navigation are fully understood there will, from desire or necessity, come the idea of building a permanent home for men in space. The ease of actual navigation in space together with the difficulties of taking-off from or landing on planets like the earth with considerable gravitational fields will in the first place lead to the necessity for bases for repairs and supplies not involving these difficulties. A damaged space vessel would, for instance, almost be bound to be destroyed in attempting earth landing. At first space navigators, and then scientists whose observations would be best conducted outside the earth, and then finally those who for any reason were dissatisfied with earthly conditions would come to inhabit these bases and found permanent spatial colonies. Even with our present primitive knowledge we can plan out such a celestial station in considerable detail.

Imagine a spherical shell ten miles or so in diameter, made of the lightest materials and mostly hollow; for this purpose the new molecular materials would be admirably suited. Owing to the absence of gravitation its construction would not be an engineering feat of any magnitude. The source of the material out of which this would be made would only be in small part drawn from the earth; for the great bulk of the structure would be made out of the substance of one or more smaller asteroids, rings of Saturn or other planetary detritus. The initial stages of construction are the most difficult to imagine. They will probably consist of attaching an asteroid of some hundred years or so diameter to a space vessel, hollowing it out and using the removed material to build the first protective shell. Afterwards the shell could be re-worked, bit by bit, using elaborated and more suitable substances and at the same time increasing its size by diminishing its thickness. The globe would fulfil all the functions by which our earth manages to support life. In default of a gravitational field it has, perforce, to keep its atmosphere and the greater portion of its life inside; but as all its nourishment comes in the form of energy through its outer surface it would be forced to resemble on the whole an enormously complicated single-celled plant.

The outermost layer would have a protective and assimilative character. The presence of meteoric matter in the solar system moving at high speeds in eccentric orbits would be the most formidable danger in space travelling and space inhabitation. Certain meteorite swarms could be avoided

16

altogether by keeping out of their tracks; larger meteorites could be detected at a distance by visual observation or by the effect of their gravitational fields. These might be avoided by changing the course of the globe or deflecting the meteorites by firing high-speed projectiles into them. Smaller meteorites would be impossible to avoid. The shell of the globe would have to be made strong enough not to be penetrated or cracked by them, and would have to possess regenerative mechanisms for repairing superficial damage. Possibly the function which our atmosphere performs for the earth could be imitated by jets of high-speed gas or electrons which, projected at meteorites, would vaporize them and thus prevent them doing any damage. At the same time meteoric matter might be the chief source of the material required for the growth or propulsion of the globe if a method of assimilating it could be found.

The outer shell would be hard, transparent and thin. Its chief function would be to prevent the escape of gases from the interior, to preserve the rigidity of the structure, and to allow the free access of radiant energy. Immediately underneath this epidermis would be the apparatus for utilizing this energy either in the form of a network carrying a chlorophyll-like fluid capable of re-synthesizing carbohydrate bodies from carbon dioxide,. or some purely electrical contrivance for the absorption of radiant energy. In the latter case the globe would almost certainly be supplied with vast, tenuous, membranous wings which would increase its area of utilization of sunlight. The subcutaneous circulation would also have the necessary function of dissipating superfluous heat, in as low temperature radiation as possible. Underneath this layer would probably lie the main stores of the globe in the form of layers of solid oxygen, ice and carbon or hydro-carbons. Inside these layers, which might be a quarter of a mile in thickness, would lie the controlling mechanisms of the globe. These mechanisms would primarily maintain the general metabolism, that is, they would regulate the atmosphere and climate both as to composition and movements. They would elaborate the necessary food products and distribute mechanical energy where it was required. They would also deal with all waste matters, reconverting them with the use of energy into a consumable form; for it must be remembered that the globe takes the place of the whole earth and not of any part of it, and in the earth nothing can afford to be permanently wasted. In this layer, too, would be the workshops and laboratories concerned with the improvement of the globe and arrangements for its growth.

Inside the mechanical layer would be the living region and here imagination has a more difficult task. It would, of course, not be necessary to have either houses or rooms in the same sense in which we have them on the earth. The absence of bad weather and of gravitation makes most of the uses that we have for houses superfluous. Perhaps we can safely assume that a certain number of cells closed by thin, but sound-proof, partitions would be necessary for work requiring special isolation, but the major part of the lives of the inhabitants of the globe would be spent in the free space which would occupy the greater portion of the center of the globe.

This three-dimensional, gravitationless way of living is very difficult for us to imagine, but there is no reason to suppose that we would not ultimately adjust ourselves to it. We should be released from the way we are dragged down on the surface of the earth all our lives: the slightest push against a relatively rigid object would send us yards away; a good jump - and we should be spinning across from one side of the globe to the other. Resistance to the air would, of course, come in, as it does on earth; but this could be turned to advantage by the use of short wings. Objects would become endowed with a peculiar levity. We should have to devise ways of holding them in place other than by putting them down; liquids and powders would at first cause great complications. An attempt to put down a cup of tea would result in the cup descending and the tea remaining as a vibrating globule in the air. Dust would be an unbearable nuisance and would have to be suppressed, because even wetting it would never make it settle. We should find in the end that all these things were great conveniences, but at first they would be extremely awkward. The possibilities of three-dimensional life would make the globes much roomier than their size would suggest. A globe interior eight miles across would contain as much effective space as a countryside one hundred and fifty miles square even if one gave a liberal allowance of air, say fifty feet above the ground.

The activity of the globe is, of course, by no means confined to its interior. In the first place it would necessarily have a number of effective sense and motor organs. Essentially the former would consist of an observatory which continually recorded the position of the globe and at the same time kept a look-out for an meteoric bodies of perceptible size which might damage it. On the whole the globe would not be designed for travel. It would move in an orbit around the sun without any expenditure of energy; but occasionally it might be necessary to shift its orbital position to a more advantageous one, and for this it would require a small motor of a rocket variety.

Yet the globe would be by no means isolated. It would be in continuous communication by wireless with other globes and with the earth, and this communication would include the transmission of every sort of sense message which we have at present acquired as well as those which we may require in the future. Interplanetary vessels would insure the transport of men and materials, and see to it that the colonies were not isolated units.

However, the essential positive activity of the globe or colony would be in the development, growth and reproduction of the globe. A globe which was merely a satisfactory way of continuing life indefinitely would barely be more than a reproduction of terrestrial conditions in a more restricted sphere. But the necessity of preserving the outer shell would prevent a continuous alteration of structure, and development would have to proceed either by the crustacean-like development in which a

19

new and better globe could be put together inside the larger one, which could be subsequently broken open and re-absorbed; or, as in the molluscs, by the building out of new sections in a spiral form; or, more probably, by keeping the even simpler form of behavior of the protozoa by the building of a new globe outside the original globe, but in contact with it until it should be in a position to set up an independent existence.

So far we have considered the construction and mechanism of the globe rather than its inhabitants. The inhabitants can be divided into the personnel or the crew, and the citizens or passengers. With the first - except that their tasks would be more complicated and more scientific than those that fall to the officers and crew of a modern ship - we need not be concerned. To the others the globe would appear both as hotels and laboratories. The population of each globe would be by no means fixed; constant interchange would be taking place between them and the earth even when the greater portion of human beings were actually inhabiting globes. There would probably be no more need for government than in a modern hotel: there would be a few restrictions concerned with the safety of the vessel and that would be all.

Criticism might be made on the ground that life in a globe, say of twenty or thirty thousand inhabitants would be extremely dull, and that the diversity of scene, of animals and plants and historical associations which exist even in the smallest and most isolated country on earth would be lacking. This criticism is valid on the initial assumption that men have not in any way changed. Here, to make globe life plausible, we must anticipate the later chapters and assume men's interests and occupations to have altered. Already the scientist is more immersed in his work and concentrates more on relations with his colleagues than in the immediate life of his neighborhood. On the other hand, present æsthetic tendencies verge towards the abstract and do not demand so much inspiration from untouched nature. What has made a small town or a small country seem in the past a

20

narrow sphere of interest has been on the one hand its isolation, and on the other hand the fact that the majority of its inhabitants are at so low a level of culture as to prevent any considerable intellectual interchange within its boundaries. Neither limitation holds for the globes, and the case of ancient Athens is enough to show that small size alone does not prevent cultural activity. Free communications and voluntary associations of interested persons will be the rule, and for those whose primary interest is in primitive nature there will always remain the earth which, free from the economic necessity of producing vast quantities of agricultural products, could be allowed to revert to a very much more natural state.

As the globes multiplied they would undoubtedly develop very differently according to their construction and to the tendencies of their colonists, and at the same time they would compete increasingly both for the sunlight which kept them alive and for the asteroidal and meteoric matter which enabled them to grow. Sooner or later this pressure, or perhaps the knowledge of the imminent failure of the sun, would force some more adventurous colony to set out beyond the bounds of the solar system. The difficulty involved in making this jump is probably as great as that of leaving the earth itself. Interstellar distances are so large that high velocities, approaching those of light, would be necessary; and though high velocities would be easy to attain - it being merely a matter of allowing acceleration to accumulate - they would expose the space vessels to very serious dangers, particularly from dispersed meteoric bodies. A space vessel would, in fact, have to be a comet, ejecting from its anterior end a stream of gas which, meeting and vaporizing any matter in its path, would sweep it to the sides and behind in a luminous trail. Such a method would be very wasteful of matter, and one might perhaps count on some better one having been devised by that time. Even with such velocities journeys would have to last for hundreds and thousands of years, and it would be necessary - if man remains as he is - for colonies of ancestors to start out who might expect the arrival of remote descendants. This would require a self-sacrifice and a perfection of educational method that we could hardly demand at the present. However, once acclimatized to space living, it is unlikely that man will stop until he has roamed over and colonized most of the sidereal universe, or that even this will be the end. Man will not ultimately be content to be parasitic on the stars but will invade them and organize them for his own purposes.

A star is essentially an immense reservoir of energy which is being dissipated as rapidly as its bulk will allow. It may be that, in the future, man will have no use for energy and be indifferent to stars except as spectacles, but if (and this seems more probable) energy is still needed, the stars cannot be allowed to continue to in their old way, but will be turned into efficient heat engines. The second law of thermodynamics, as Jeans delights in pointing out to us, will ultimately bring this universe to an inglorious close, may perhaps always remain the final factor. But by intelligent organization the life of the universe could probably be prolonged to many millions of millions of times what it would be without organization. Besides, we are still too close to the birth of the universe to be certain about its death. In any case, long before these questions become urgent it would seem impossible not to assume that man himself would have changed radically in this environment and the nature of this change we must consider in the next chapter.

III. The Flesh

In the alteration of himself man has a great deal further to go than in the alteration of his inorganic environment. He has been doing the latter more or less unconsciously and empirically for several thousand years, ever since he cased being parasitic on his environment like any other animal, and consciously and intelligently for at least hundreds of years; whereas he has not been able to change himself at all and has had only fifty years or so to begin to understand how he works. Of course, this is not strictly true: man has altered himself in the evolutionary process, he has lost a good deal of hair, his wisdom teeth are failing to pierce, and his nasal passages are becoming more and more degenerate. But the processes of natural evolution are so much slower than the development of man's control over environment that we might, in such a developing world, still consider man's body as constant and unchanging. If it is not to be so then man himself must actively interfere in his own making and interfere in a highly unnatural manner. The eugenists and apostles of healthy life, may, in a very considerable course of time, realize the full potentialities of the species: we may count on beautiful, healthy and long-lived men and women, but they do not touch the alteration of the species. To do this we must alter either the germ plasm or the living structure of the body, or both together. The first method - the favorite of Mr. J. B. S. Haldane - has so far received the most attention. With it we might achieve such a variation as we have empirically produced in dogs and goldfish, or perhaps even manage to produce new species with special potentialities. But the method is bound to be slow and finally limited by the possibilities of flesh and blood. The germ plasm is a very inaccessible unit, before we can deal with it adequately we must isolate it, and to do this already involves us in surgery. It is quite conceivable that the mechanism of evolution, as we know it up to the present, may well be superseded at this point. Biologists are apt, even if they are not vitalists, to consider it as almost divine; but after all it is only nature's way of achieving a shifting equilibrium with an environment; and if we can find a more direct way by the use of intelligence, that way is bound to supersede the unconscious mechanism of growth and reproduction.

In a sense we have already started using the direct method; when the ape-ancestor first used a stone he was modifying his bodily structure by the inclusion of a foreign substance. This inclusion was temporary, but with the adoption of clothes there began a series of permanent additions to the body, affecting nearly all its functions and even, as with spectacles, its sense organs. In the modern world, the variety of objects which really form part of an effective human body is very great. Yet they all (if we except such rarities as artificial larynges) still have the quality of being outside the cell layers of the human body. The decisive step will come when we extend the foreign body into the actual structure of living matter. Parallel with this development is the alteration of the body by tampering with its chemical reactions - again a very old-established but rather sporadic process resorted to to cure illness or procure intoxication. But with the development of surgery on the one hand and physiological chemistry on the other, the possibility of radical alteration of the body appears for the first time. Here we may proceed, not by allowing evolution to work the changes, but by copying and short-circuiting its methods.

The changes that evolution produces apart from mere growth in size, or diversity of form without change of function, are in the nature of perversions: a part of the fish's gut becomes a swimming bladder, the swimming bladder becomes a lung; a salivary gland and an extra eye are charged with the function of producing hormones. Under the pressure of environment or whatever else is the cause of evolution, nature takes hold of what already had existed for some now superseded activity, and with a minimum of alteration gives it a new function. There is nothing essentially mysterious in the process: it is both the easiest and the only possible way of achieving the change. Starting *de novo* to deal with a new situation is not within the power of natural, unintelligent processes; they can only modify in a limited way already existing structures by altering their chemical environment. Men may well copy the process, in so far as original structures are used as the basis for new ones, simply because it is the most economical method, but they are not bound to the very limited range of methods of change which nature adopts.

Now modern mechanical and modern chemical discoveries have rendered both the skeletal and metabolic functions of the body to a large extent useless. In teleological biochemistry one might say that an animal moves his limbs in order to get his food, and uses his body organs in order to turn that food into blood to keep his body alive and active. Now if man is only an animal this is all very satisfactory, but viewed from the standpoint of the mental activity by which he increasingly lives, it is a highly inefficient way of keeping his mind working. In a civilized worker the limbs are mere parasites, demanding nine-tenths of the energy of the food and even a kind of blackmail in the exercise they need to prevent disease, while the bodily organs wear themselves out in supplying their requirements. On the other hand, the increasing complexity of man's existence, particularly the mental capacity required to deal with its mechanical and physical complications, gives rise to the need for a much more complex sensory and motor organization, and even more fundamentally for a better organized cerebral mechanism. Sooner or later the useless parts of the body must be given more modern functions or dispensed with altogether, and in their place we must incorporate in the effective body the mechanisms of the new functions. Surgery and biochemistry are sciences still too young to predict exactly how this will happen. The account I am about to give must be taken rather as a fable.

Take, as a starting point, the perfect man such as the doctors, the eugenists and the public health officers between them hope to make of humanity: a man living perhaps an average of a hundred and twenty years but still mortal, and increasingly feeling the burden of this mortality. Already Shaw in his mystical fashion cries out for life to give us hundreds of years to experience, learn and understand; but without the vitalist's faith in the efficacy of human will we shall have to resort to some artifice in order to achieve this purpose. Sooner or later some eminent physiologist will have his neck broken in a super-civilized accident or find his body cells worn beyond capacity for repair.

He will then be forced to decide whether to abandon his body or his life. After all it is brain that counts, and to have a brain suffused by fresh and correctly prescribed blood is to be alive - to think. The experiment is not impossible; it has already been performed on a dog and that is three-quarters of the way towards achieving it with a human subject. But only a Brahmin philosopher would care to exist as an isolated brain, perpetually centered on its own meditations. Permanently to break off all communications with the world is as good as to be dead. However, the channels of communication are ready to hand. Already we know the essential electrical nature of nerve impulses; it is a matter of delicate surgery to attach nerves permanently to apparatus which will either send messages to the nerves or receive them. And the brain thus connected up continues an existence, purely mental and with very different delights from those of the body, but even now perhaps preferable to complete extinction. The example may have been too far-fetched; perhaps the same result may be achieved much more gradually by use of the many superfluous nerves with which our body is endowed for various auxiliary and motor services. We badly need a small sense organ for detecting wireless frequencies, eyes for infra-red, ultra-violet and X-rays, ears for supersonics, detectors of high and low temperatures, of electrical potential and current, and chemical organs of many kinds. We may perhaps be able to train a great number of hot and cold and pain receiving nerves to take over these functions; on the motor side we shall soon be, if we are not already, obliged to control mechanisms for which two hands and feet are an entirely inadequate number; and, apart from that, the direction of mechanism by pure volition would enormously simplify its operation. Where the motor mechanism is not primarily electrical, it might be simpler and more effective to use nerve-muscle preparations instead of direct nerve connections. Even the pain nerves may be pressed into service to report any failure in the associated mechanism. A mechanical stage, utilizing some or all of these alterations of the bodily form might, if the initial experiments were successful in the sense of leading to a tolerable existence, become the regular culmination to ordinary life. Whether this should ever be so for the whole of the population we will discuss later, but for the moment we may attempt to picture what would at this period be the course of existence for a transformable human being.

Starting, as Mr. J. B. S. Haldane so convincingly predicts, in an ectogenetic factory, man will have anything from sixty to a hundred and twenty years of larval, unspecialized existence - surely enough to satisfy the advocates of a natural life. In this stage he need not be cursed by the age of science and mechanism, but can occupy his time (without the conscience of wasting it) in dancing, poetry and love-making, and perhaps incidentally take part in the reproductive activity. Then he will leave the body whose potentialities he should have sufficiently explored.

The next stage might be compared to that of a chrysalis, a complicated and rather unpleasant process of transforming the already existing organs and grafting on all the new sensory and motor mechanisms. There would follow a period of re-education in which he would grow to understand the functioning of his new sensory organs and practise the manipulation of his new motor mechanism. Finally, he would emerge as a completely effective, mentally-directed mechanism, and set about the tasks appropriate to his new capacities. But this is by no means the end of his development, although it marks his last great metamorphosis. Apart from such mental development as his increased faculties will demand from him, he will be physically plastic in a way quite transcending the capacities of untransformed humanity. Should he need a new sense organ or have a new mechanism to operate, he will have undifferentiated nerve connections to attach to them, and will be able to extend indefinitely his possible sensations and actions by using successively different end-organs.

The carrying out of these complicated surgical and physiological operations would be in the hands of a medical profession which would be bound to come rapidly under the control of transformed men. The operations themselves would probably be conducted by mechanisms controlled by the transformed heads of the profession, though in the earlier and experimental stages, of course, it would still be done by human surgeons and physiologists.

It is much more difficult to form a picture of the final state, partly because this final state would be so fluid and so liable to improve, and partly because there would be no reason whatever why all people should transform in the same way. Probably a great number of typical forms would be developed, each specialized in certain directions. If we confine ourselves to what might be called the first stage of mechanized humanity and to a person mechanized for scientific rather than æsthetic purposes - for to predict even the shapes that men would adopt if they would make of *themselves* a harmony of form and sensation must be beyond imagination - then the description might run roughly as follows.

Instead of the present body structure we should have the whole framework of some very rigid material, probably not metal but one of the new fibrous substances. In shape it might well be rather a short cylinder. Inside the cylinder, and supported very carefully to prevent shock, is the brain with its nerve connections, immersed in a liquid of the nature of cerebro-spinal fluid, kept circulating over it at a uniform temperature. The brain and nerve cells are kept supplied with fresh oxygenated blood and drained of de-oxygenated blood through their arteries and veins which connect outside the cylinder to the artificial heart-lung digestive system - an elaborate, automatic contrivance. This might in large part be made from living organs, although these would have to be carefully arranged so that no failure on their part would endanger the blood supply to the brain (only a fraction of the body's present requirements) and so that they could be inter-changed and repaired without disturbing its functions. The brain thus guaranteed continuous awareness, is connected in the anterior of the case with its immediate sense organs, the eye and the ear - which will probably retain this connection for a long time. The

28

eyes will look into a kind of optical box which will enable them alternatively to look into periscopes projecting from the case, telescopes, microscopes and a whole range of televisual apparatus. The ear would have the corresponding microphone attachments and would still be the chief organ for wireless reception. Smell and taste organs, on the other hand, would be prolonged into connections outside the case and would be changed into chemical tasting organs, achieving a more conscious and less purely emotional role than they have at present. It may perhaps be impossible to do this owing to the particularly close relation between the brain and olfactory organs, in which case the chemical sense would have to be indirect. The remaining sensory nerves, those of touch, temperature, muscular position and visceral functioning, would go to the corresponding part of the exterior machinery or to the blood supplying organs. Attached to the brain cylinder would be its immediate motor organs, corresponding to but much more complex than, our mouth, tongue and hands. This appendage system would probably be built up like that of a crustacean which uses the same general type for antenna, jaw and limb; and they would range from delicate micro-manipulators to lever capable of exerting considerable forces, all controlled by the appropriate motor nerves. Closely associated with the brain-case would also be sound, color and wireless producing organs. In addition to these there would be certain organs of a type we do not possess at present - the self-repairing organs - which under the control of the brain would be able to manipulate the other organs, particularly the visceral blood supply organs, and to keep them in effective working order. Serious derangements, such as those involving loss of consciousness would still, of course, call for outside assistance, but with proper care these would be in the nature of rare accidents.

The remaining organs would have a more temporary connection with the brain-case. There would be locomotor apparatus of different kinds, which could be used alternatively for slow movement, equivalent to walking, for rapid transit and for flight. On the whole, however, the locomotor organs would not be much used because the extension of the sense organs would tend to take their place. Most of these would be mere mechanisms quite apart from the body; there would be the sending parts of the television apparatus, tele-acoustic and tele-chemical organs, and tele-sensory organs of the nature of touch for determining all forms of textures. Besides these there would be various tele-motor organs for manipulating materials at great distances from the controlling mind. These extended organs would only belong in a loose sense to any particular person, or rather, they would belong only temporarily to the person who was using them and could equivalently be operated by other people. This capacity for indefinite extension might in the end lead to the relative fixity of the different brains; and this would, in itself, be an advantage from the point of view of security and uniformity of conditions, only some of the more active considering it necessary to be on the spot to observe and do things.

The new man must appear to those who have not contemplated him before as a strange, monstrous and inhuman creature, but he is only the logical outcome of the type of humanity that exists at present. It may be argued that this tampering with bodily mechanisms is as unnecessary as it is difficult, that all the increase of control needed may be obtained by extremely responsive mechanisms outside the unaltered human body. But though it is possible that in the early stages a surgically transformed man would be at a disadvantage in capacity of performance to a normal, healthy man, he would still be better off than a dead man. Although it is possible that man has far to go before his inherent physiological and psychological make-up becomes the limiting factor to his development, this must happen sooner or later, and it is then that the mechanized man will begin to show a definite advantage. Normal man is an evolutionary dead end; mechanical man, apparently a break in organic evolution, is actually more in the true tradition of a further evolution.

A much more fundamental break is implicit in the means of his development. If a method has been found of connecting a nerve ending in a brain directly with an electrical reactor, then the way is open for connecting it with a brain-cell of another person. Such a connection being, of course, essentially electrical, could be effected just as well through the ether as along wires. At first this would limit itself to the more perfect and economic transference of thought which would be necessary in the co-operative thinking of the future. But it cannot stop here. Connections between two or more minds would tend to become a more and more permanent condition until they functioned as a dual or multiple organism. The minds would always preserve a certain individuality, the network of cells inside a single brain being more dense than that existing between brains, each brain being chiefly occupied with its individual mental development and only communicating with the others for some common purpose. Once the more or less permanent compound brain came into existence two of the ineluctable limitations of present existence would be surmounted. In the first place death would take on a different and far less terrible aspect. Death would still exist for the mentally-directed mechanism we have just described; it would merely be postponed for three hundred or perhaps a thousand years, as long as the brain cells could be persuaded to live in the most favorable environment, but not forever. But the multiple individual would be, barring cataclysmic accidents, immortal, the older component as they died being replaced by newer ones without losing the continuity of the self, the memories and feelings of the older member transferring themselves almost completely to the common stock before its death. And if this seems only a way of cheating death, we must realize that the individual brain will feel itself part of the whole in a way that completely transcends the devotion of the most fanatical adherent of a religious sect. It is admittedly difficult to imagine this state of affairs effectively. It would be a state of ecstasy in the literal sense, and this is the second great alteration that the compound mind makes possible. Whatever the intensity of our feeling, however much we may strive to reach beyond ourselves or into another's mind, we are always barred by the limitations of our individuality. Here at least those barriers would be down: feeling would truly communicate itself, memories would be held in common, and yet in all this, identity and continuity of individual development would not be lost. It is possible, even probably, that the different individuals of a compound mind would not all have similar functions or even be of the same rank of importance. Division of labor would soon set in: to some minds might be delegated the task of ensuring the proper functioning of the others, some might specialize in sense reception and so on. Thus would grow up a hierarchy of minds that would be more truly a complex than a compound mind.

The complex minds could, with their lease of life, extend their perceptions and understanding and their actions far beyond those of the individual. Time senses could be altered: the events that moved with the slowness of geological ages would be apprehended as movement, and at the same time the most rapid vibrations of the physical world could be separated. As we have seen, sense organs would tend to be less and less attached to bodies, and the host of subsidiary, purely mechanical agents and preceptors would be capable of penetrating those regions where organic bodies cannot enter or hope to survive. The interior of the earth and the stars, the inmost cells of living things themselves, would be open to consciousness through these angels, and through these angels also the motions of stars and living things could be directed.

This is perhaps far enough; beyond that the future must direct itself. Yet why should we stop until our imaginations are exhausted. Even beyond this there are foreseeable possibilities. Undoubtedly the nature of life processes themselves will be far more intensively studied. To make life itself will be only a preliminary stage, because in its simplest phases life can differ very little from the inorganic world. But the mere making of life would only be important if we intended to allow it to evolve of itself anew. This, as Mr. Whyte suggests in *Archimedes,* is necessarily a lengthy process, but there is no need to wait for it. Instead, artificial life would undoubtedly be used as ancillary to human activity and not allowed to evolve freely except for experimental purposes. Men will not be content to manufacture life: they will want to improve on it. For one material out of which nature has been forced to make life, man will have a thousand; living and organized material will be as much at the call of the mechanized or compound man as metals are to-day, and gradually this living material will come to substitute more and more for such inferior functions of the brain as memory, reflex actions, etc., in the compound man himself; for bodies at this time would be left far behind. The brain itself would become more and more separated into different groups of cells or individual cells with complicated connections, and probably occupying considerable space. This would mean loss of motility which would not be a disadvantage owing to the extension of the sense faculties. Every part would not be accessible for replacing or repairing and this would in itself ensure a practical eternity of existence, for even the replacement of a previously organic brain-cell by a synthetic apparatus would not destroy the continuity of consciousness.

The new life would be more plastic, more directly controllable and at the same time more variable and more permanent than that produced by the triumphant opportunism of nature. Bit by bit the heritage of the direct line of mankind - the heritage of the original life emerging on the face of the world - would dwindle, and in the end disappear effectively, being preserved perhaps as some curious relic, while the new life which conserves none of the substance and all of the spirit of the old would take its place and continue its development. Such a change would be as important as that in which life first appeared on the earth's surface and might be as gradual and imperceptible.

Finally, consciousness itself may end or vanish in a humanity that has become completely etherealized, losing the close-knit organism, becoming masses of atoms in space communicating by radiation, and ultimately perhaps resolving itself entirely into light. That may be an end or a beginning, but from here it is out of sight.

IV. Devil

Why do the first lines of attack against the inorganic forces of the world and the organic structure of our bodies seem so doubtful, fanciful and Utopian? Because we can abandon the world and subdue the flesh only if we first expel the devil, and the devil, for all that he has lost individuality, is still as powerful as ever. The devil is the most difficult of all to deal with: he is inside ourselves, we cannot see him. Our capacities, our desires, our inner confusions are almost impossible to understand or cope with in the present, still less can we predict what will be the future of them. Psychology at the present day is hardly in a better state than physics in the time of Aristotle; it has acquired a vocabulary, the general movements and transformations of conscious and unconscious motives are described, but nothing more. Yet in the absence of scientific analysis something must be said, because all the changes I have predicted in the organic or inorganic world must, in the first place, start from some human psychological motive and effect themselves through the operation of human intellectual processes. We are obviously not in a position to predict the particular new orientations which a change in psychology would give to human development, beyond that which would result from the removal of what we know are inhibitory causes, so that here I will only attempt to estimate the effect of psychological forces in preventing or retarding the kind of processes outlined in the first two sections. The progress of the future depends no longer on physiological evolution but on the reaction of intelligence on a material universe. It will be hindered or stopped either by a failure in the capacity for maintaining creative intellectual thinking, or by the lack of desire to apply such thinking to the progress of humanity, or, of course, by both these causes together. Consider first the retarding factors that endanger the capacity for creative thinking. Some are apparent now. It is pretty clear that they are ineffective in stopping the course of thought at present, but they have not always been so in the past and we cannot be sure that they will not be so in the future. One of the most threatening retarding factors of the present is specialization, particularly as it is bound to increase with scientific knowledge itself. But it is doubtful whether specialization in itself is capable of bringing scientific thought to a standstill. It retards it in so far as the specialist is ignorant of current thought in other fields, and the remedy for this is obviously an intelligently operated system of distribution and grading of knowledge so that each worker may have the amount he requires outside his own field, in a form which can be absorbed with a minimum of mental effort. The problem is essentially that of communications to an army in action. After a rapid advance communications become disorganized, and there is a temporary halting until they are again in working order.

Such an organization of intellectual work for definite ends involves a fundamental change: it is analogous to the change from a good-gathering to a food-producing society. The modern scientist is a primitive savage. If he is active and enterprising he tracks his prey down alone or in small parties; if he is industrious and thorough he gathers and piles up the natural products around him, but for his success he has to thank not only his own skill and the lore of his craft but the richness of nature and the paucity of his companions. Good hunting will not last much longer, but the tilled ground is richer.

We shall be forced to attempt planned and directed research employing hundreds of workers for many years, and this cannot be done without risking the loss of independence and originality. This is a serious and fundamental obstacle but it may be overcome in two ways. It should be possible so to improve educational methods, that mental activity, the capacity to form new associations, should not be incompatible with the performance of routine work: that is, every research worker should be potentially able to add to and modify the whole course of the research and suggestions. At the same time it is certain that originality, organizing power and industriousness will continue as now to be very unevenly distributed; and it is an essentially social problem to make the best apportionment of functions, using for the more routine operations people who under present conditions would not be scientific workers at all, and using the organizers to translate into plans of action the incoherent ideas of the thinkers. Pedantry and bureaucracy - symptoms of an unintelligent respect for the past - are at present real dangers, but, once their genesis is understood, they can be made to vanish.

Specialization is brought about by the wideness of the field in which science operates, but as we go more deeply into nature the intrinsic complication of the phenomena increases and the modes of thinking used in ordinary life become more inadequate to deal with them. It is conceivable that the supply of minds capable of making any impression on these deeper problems may more and more fall behind the number required, and that all the efforts of education to produce ten genii where one grew before will be foiled by intrinsic difficulties in nature. It is impossible to know whether this will happen. One may guess, from experience of the past, that nature is never so complicated as it looks; that the value of theory and deductive thinking and the use of appropriate language and symbolism will reduce the difficulties in the measure that they are approached.

However they appear to the pessimist of the present day, it is not in specialization or complication that the chief danger to progress seems to lie: it is in something much more deep-seated and much more elusive. Bertrand Russell, in one of his *Skeptical Essays,* predicting the approaching end of the scientific age, suggests that people will turn from physics to metaphysics because the hope that the former held out is seen to be vain except to new, half-cultivated peoples. Perhaps after all it is hope that really determines whether an age is or is not creative. But the existence of hope in a society at any time itself depends on many unexplored psychological, economic and political causes. I do not think that the factors involved are of a mystical order, but that they require considerable disentangling.

There seem to be two psychological determinants in any culture: a crop of perverted individuals capable of more than average performance, and a mass of people effective not so much by their number as by their secure hold on tradition. In the normal state the perverse are dominated by the mass in two ways. Their mode of expression is dictated by the modes conceivable in the society; everywhere, even the most aberrant individual must conform to one of a small number of recognized types. The same type of mind that would now make a physicist would in the middle ages have made a scholastic theologian. Further, there is a process of selection in which the current tradition decides what is to be the relative value and effectiveness of each type. Thus, even, if at all times types are always produced in the same abundance, only the selected are effective, as meditative ascetics in India or energetic salesmen in America. The mass of the people, or more properly the ruling class, pay the piper and call the tune; genius is potent only when it fits the tendencies of the age. From this standpoint we are approaching the close of the period of respectable comfort which puritanism demanded and mathematics and handicraft produced. But this period may not end in a regression to the mediæval state through the ultimate dissatisfaction with science; before that happens science, raised to power by industrialism, may in its turn become the directing tradition.

Political and social events must also be effective, but not in a very obvious fashion. But political confusion and prolonged peace undoubtedly affect creative thought but whether they respectively hinder or help it is not at all certain. When one contrasts Athens, renaissance Italy and feudal China, on the one hand, with the Roman, the Spanish and the Chinese Empires on the other, war would seem positively to help mental activity. But as many examples could be found to the contrary. There may be something in the suggestion that wherever war appeared stimulating it was a war between approximate equals so that the disasters were seen to be due to human folly or perversity. In the case of of the Empires, on the other hand, peace was achieved at the price of a submission to authority, bureaucratic or spiritual, which deprived men of their self-reliance and creative ability. However this may be, historical factors tend to have somewhat of a cyclic nature, and in the long run to cancel each other out, although it is always possible that one age will destroy, or cause to be forgotten, more than the previous ages produced, and that a definite culmination may be reached in human progress. This may be closer than we think (if it is not already passed) and humanity may become static until it is destroyed by cosmic forces. Yet it seems more probably that we are on the point, owing to our material achievement of reaching another order of cyclic changes, which may lead us to the stars.

Whether an age or an individual will express itself in creative thinking or in repetitive pedantry is more a matter of desire than of intellectual power, and it is probably more the nature of their desires than of their capacities that will determine whether or not humanity will develop further. Now it would seem that the present time is a very critical one for the evolution of human desire. It is an age in which the nature of desire has been glimpsed at for the first time, and that glimpse enables us to see two very different possibilities. The intellectual life, both in its scientific and its æsthetic aspects, is seen no longer as the vocation of the rational mind, but as a compensation, as a perversion of more primitive, unsatisfied desires. Now the question arises is this perversion in the line of evolution, or is it a merely temporary, pathological process? If by a sounder psychology, a way of living more in accordance with nature, it should be found that the satisfaction of purely human - or, as we might almost say, purely mammalian - desires is capable of absorbing all the energy that suppression now forces into scientific or æsthetic channels, then the human race may well find itself statically employed in leading an idyllic, Melanesian existence of eating, drinking, friendliness, love-making, dancing and singing, and the golden age may settle permanently on the world. On the other hand it may that though the desire, the necessity to escape life on the paths of intellectual or æsthetic creation may be weakened by the application of an intelligent psychology, yet a corresponding freedom from the internal conflicts which now hinder both these forms of expression may more than compensate for what is lost, and we may find the capacity to live at the same time more fully human and fully intellectual lives. The latter alternative is more in line with the recent developments of Freudian psychology which divide the psyche into the primitive id, the ego which is its expression of contact with reality, and the super-ego which represents its aspirations and ideals. Rationalism strove to make the super-ego the dominant partner; it never succeeded, not only because its standard was too high to allow any outlet for the primitive forces, but because it was itself too arbitrary, too tainted with distorted primitive wishes ever to be brought into correspondence with reality. Naturalism, less definitely, aimed at giving the primitive wishes full play but equally failed because these wishes are too primitive, too infantile, too inconsistent with themselves to be satisfied even by the greatest license. The aim of applied psychology is now to bring, by analysis or education, the ideals of the super-ego in line with external reality, using and rendering innocuous the power of the id and leading to a life where a full adult sexuality would be balanced with objective activity. It is this alternative that makes the mechanical, biological progress that I have outlined not only possible but almost necessary, for a sound intellectual humanity will never be content with repeating itself in circles of metaphysical thinking like Shaw's Immortals, but will need a real externalization in the transforming the universe and itself. Such a development could hardly leave unchanged the present types of human interests in art and science and religion.

It is here that prediction is most difficult and most fascinating. Under the influence of psychology it may well be that, just as all the branches of science itself are coalescing into a unified world picture, so the human activities of art and attitudes of religion may be fused into one whole action-reaction pattern of man to reality. The recognition of the art that informs all pure science need not mean the abandonment for it of all present art, rather it will mean the completion of the transformation of art that has already begun. Art expressing itself on one side in a kind of generalized architecture, massive or molecular, gives form to the infinite possibilities of the application of science; on the other a generalized poetry expresses the ever-widening complexities of the understanding of the universe, while religion clarified by psychology remains at the expression of the desire that drives man through the universe in understanding and hope.

It is not sufficient, however, to consider the absence or presence of desire for progress, because that desire itself will not make itself effective until it can overcome the quite real distaste and hatred which mechanization has already brought into being. This distaste is nothing to what the bulk of present humanity would feel about even the milder of the changes which are suggested here. The reader may have already felt that distaste, especially in relation to the bodily changes; I have felt it myself in imagining them. The effectiveness of these conservative feelings is the balance of two opposing factors. The changes in question do not come all at once: envisaged in broad outline in the sequence given, their nature would suggest that they follow each other with increasing frequency, as the past has already shown. Now the more rapid the environmental changes the less will the individual mind be able to adapt itself to them and the more violent will be its emotional reactions. At the same time these changes give more and more power to those groups of men which are involved in them and are bringing them about, so that, up to the present, in the war of the machines, the mechanists have always been the victors; but, of course, if the emotional reactions of the mass increased more rapidly than the power of the mechanists, the reverse would be the case. A severe crisis in mechanical civilization brought about by its inherent technical weakness or, as is much more likely, by its failure to arrange secondary social adjustments, is likely to be seized upon by the emotional factors hostile to all mechanism, and we may be closer to such a reversion than we suppose. To recent books representing very divergent standpoints, the last works of Mr. Aldous Huxley and Mr. D. H. Lawrence, show at the same time the weakening desires and the imminent realization of futility on the part of the scientist, and a turning away from the whole of mechanization on the part of the more humanely-minded. The same thought is echoed from still another angle in the writings of Mr. Bertrand Russell. They may be prophets predicting truly the doom of the new Babylon or merely lamenting over a past that is lost for ever. With these uncertainties before us, each must follow his own desires, accepting that his opponent may be as right as himself. The event will show which, but only after his own time.

There remains still another possibility: the most unexpected, but not necessarily the most improbable, the development of a di-morphism in humanity in which the conflict between the humanizers and the mechanizers will be solved not by the victory of one or the other but by the splitting of the human race - the one section developing a fully-balanced humanity, the other groping unsteadily beyond it. But this possibility involves the consideration of mechanical and biological factors, the synthesis of which, with the psychological, will be attempted in my concluding pages.

V. Synthesis

Having followed our main lines of change separately, it

now remains for us to consider the interaction between the physical, physiological and psychological elements of future human evolution. It is very easy to see the relations of the first two: the colonization of space and the mechanization of the body are obviously complementary. The dissimilarity between the conditions of life in space and on the earth would in itself be sufficient to cause perfectly normal, unassisted, evolutionary changes in human beings, but obviously spatial conditions would be more favorable to mechanized than to organic man. If he could get rid of the major part of his body and his necessity for a relatively large intake of oxygen and water-saturated food, the cellular nature of the celestial globes would cease to be necessary. This would give mechanized man an advantage similar to that which the relatively flexible and naked animal cell has over the rigidly demarcated plant. Besides, it is only in space that the potentialities of the more highly developed forms of complex minds would have an adequate field of functioning, particularly in their extended time relations.

It may be that we are approaching or will ultimately reach a conception of time that will make transit in time as easy as transit in space. But all our present knowledge, apart from our desires, suggests that it is improbable. Even if time and space were made equivalent, to gain a second of the future would be equivalent to travelling 180,000 miles. But even without a fundamental change in the conception of time the time faculties of mechanized man would still be very different from ours. Extension will be its chief character: already in the monkey stage the actual present of an animal embraces a short part of the past and future. Anticipation of movement, through muscular innervation and memory, by its retention of nerve impulse images, extend the present to the limit of a second or so. Every time we play tennis we are prophets without knowing of the future position of the ball which is conceived of as present. In the human stage we have extended mostly backwards as memory, our immediate prevision being limited by lack of scientific knowledge. It is now rapidly increasing, but is not usually accepted as prevision because it is conscious and intellectual. However, prevision plainly tends to become more and more deductive, and, to the mechanized man, the immediately apprehended may include years or centuries of past and future.

One may picture then, these beings, nuclearly resident, so to speak, in a relatively small set of mental units, each utilizing the bare minimum of energy, connected together by a complex of ethereal intercommunication, and spreading themselves over immense areas and periods of time by means of inert sense organs which, like the field of their active operations, would be, in general, at a great distance from themselves. As the scene of life would be more the cold emptiness of space than the warm, dense atmosphere of planets, the advantage of containing no organic material at all, so as to be independent of both these conditions, would be increasingly felt.

It is when we turn to the interaction on the psychological plane that the difficulties again occur. The physical and the psychological have a mutual influence which it is very difficult at the present moment to estimate. Undoubtedly, if modern tendencies have any elements of permanency in them, a great deal of the activity of the future will be devoted to the end of a greater understanding of the universe. Humanity, or its descendants, may well be much more occupied with purely scientific research and much less with the necessity of satisfying primarily physiological and psychological needs than it is at present. This character may stamp the whole of future development, so that machinery will be organized not for production but for discovery. Indeed, the great necessity for production either of food or other articles of consumption will disappear rapidly with the progress of dehumanization. But such changes are small compared with those which would necessarily be involved by the physiological alterations which I have suggested.

The human mind evolved always in the company of the

human body, and of the animal body before it was human. The intricate connections of mind and body must exceed our imagination, as from our point of view we are peculiarly prevented from observing them. Altering in any perfectly sound physiological or surgical way the functionings of the body will certainly have secondary but far-reaching effects on the mind, and these secondary effects will be still unpredictable at the time when the physiological changes take place. But it is thoroughly in accord with both human and natural evolution that secondary changes should not be taken into account when reacting to the primary desire or stimulus: in other words, the physiological steps will probably be taken without consideration of the psychological consequences, which may, of course, wreck the whole organism, or, on the other hand, lead to an unpredictable large increase in mental grasp and efficiency. It is on account of this delicate balance between physiological and psychological factors that the future, as well as the present, will be full of dangerous turning-points and pitfalls. We shall have very sane reactionaries at all periods warning us to remain in the natural and primitive state of humanity, which is usually the last stage but one in their cultural history. But the secondary consequences of what men have already done - the reactionaries as much as any - will carry them away then as now. Obviously certain considerable psychological displacements or perversions must occur to balance the physiological perversions. The sexual instincts in particular, which still find considerable direct gratification, would be unrecognizably changed. One may assume that there is some kind of principle of psychological conservation which will prevent them, as it has prevented them up to the present, being suppressed altogether. But what will they be changed into? The solution may be an extension of sublimation, a process which is at present outside conscious control but which may not always remain so. A part of sexuality may go to research, and a much larger part must lead to æsthetic creation. The art of the future will, because of the very opportunities and materials it will have at its command, need an infinitely stronger formative impulse than it does now. The cardinal tendency of progress is the replacement of an indifferent chance environment by a deliberately created one. As time goes on, the acceptance, the appreciation, even the understanding of nature, will be less and less needed. In its place will come the need to determine the desirable form of the humanly-controlled universe which is nothing more nor less than art.

The psychology of a complex mind must differ almost as much from that of a simple, mechanized mind as its psychology would from ours; because something that must underlie and perhaps be even greater than sex is involved. By the intimate intercommunication of minds, the very existence of the ego would be impaired for the first time. Some kind of equilibrium will have to be found between each partial and corporate personality. This we can vaguely adumbrate when we think of the conflicts involved between ego and sexual impulses, the latter attempting always to break the isolation of the former and reach out to another individual or a group. If it is once possible to achieve this reaching out of feeling, the results are bound to be enormous and perhaps overwhelming. Will the corporate personalities form greater and greater complexes until there is only one intelligence, or will there be a multiplication of separate and differently-evolving complexes with resulting conflicts? Spatial considerations seem on the whole to favor the latter view, but we must allow for enormous increases in communications and in the capacity for rational conduct.

Another even deeper psychological consideration arises at this point. What is to be the future of feeling? Is it to be perverted or superseded altogether? In other words, are the mechanical or corporate men of the future to be emotional or rational? Here we have very little to guide us; we are not certain whether the comparative coldness of modern intellectualism is the effect of considerable development or of dangerous perversion. Even if we did know the answer to this it would hardly help us, since our new beings would have a different physiological balance. This balance will not be, as in us, at the mercy of the uncontrolled interactions of individual and environment. Feeling, or at any rate, feeling-tones, will almost certainly be under conscious control: a feeling-tone will be induced in order to favor the performance of a particular kind of operation. Of course, it would be excessively dangerous for human beings in their present state to have this control of their feelings. A great majority would probably be content to remain in a state of more or less ecstatic happiness, but the man of the future will probably have discovered that happiness is not an end of life. This is as far as we may go even in guessing. The psychology of the completely mechanized organism must remain a mystery.

Viewed from the standpoint of the present the carrying out of such a program of human development must seem a very pointless occupation; but it is doubtful whether the present civilization would appear to an educated Athenian as something worthy to mark the culmination of his efforts. We must not assume a static psychology and a further static knowledge. The immediate future which is our own desire, we seek; in achieving it we become different; becoming different we desire something new, so there is no staleness except when development itself has stopped. Moreover, development, even in the most refined stages, will always be a very critical process; the dangers to the whole structure of humanity and its successors will not decrease as their wisdom increases, because, knowing more and wanting more they will dare more, and in daring will risk their own destruction. But this daring, this experimentation, is really the essential quality of life.

VI. Possibility

By now it should be possible to make a picture of the general scheme of development as a unified whole, and though each part may seem plausible in detail, yet in some obscure way the total result seems unbelievable. This disbelief may be well founded, for what is suggested is not so much a fulfilling as a transformation of humanity, a setting up of what is virtually a new species or several new species, and a mode of setting up which is in itself a departure from the time-hallowed methods of evolution. Now, I believe, that this scheme is more than a bare possibility, that it, or something like it, has about an even chance of occurring; but I must justify this belief not by hypothetics of the future but by analysis of causes acting in the present. Perhaps the most fruitful way is to ask the question, ``What is the effective purpose of the human race as it now is?" We can eliminate such satisfactory answers as ``For the glory of God," because, however true, they do not differentiate humanity from other parts of creation. The answer one seeks is the historical and economic one. Human societies are recent products and, up to the present, can be essentially qualified as co-operative food-producing societies - or perhaps, to include comfort, as co-operative body-satisfying societies. They are distinguished in this way from insect societies, which are essentially, as Wheeler has pointed out, reproductive societies. True, in fulfilling the function of securing a brood, insect societies have gone far in becoming food producing units, and the complementary process in man is shown by the increased care taken over education; but devotion to children has never been the mainspring of human activity. Hunger and sex still dominate the primitive mammalian side of human existence, but at the present time it looks as if humanity were within sight of their satisfaction. Permanent plenty, no longer a Utopian dream, awaits the arrival of permanent peace. Even now, through rationalized capitalism or Soviet state planning, the problem of the production and distribution of necessaries to the primary satisfaction of all human beings, is being pushed forward with uniform and intelligent method. Stupidity and the perversity of separate interests may hold the consummation back for centuries, but it must come gradually and surely.

Now supposing this state achieved or approximating to

achievement, what is to become of humanity? Is it, like the stabilized insect societies, to settle down to an eternity of methodical enjoyment, or is there appearing, by some unforeseen chance, a new objective, a new reason for existing beyond the calls of hunger and lust? The primates, and subsequently man, developed intelligence in order to satisfy their desires in a world that was getting more and more difficult to live in. They developed it as primitive plants develop the habit of eating, or fish that of breathing, and just as those plants became animals who lived to eat and those fish became animals who lived to breathe, so we may, in time, come to live to think instead of thinking to live. But this biological analogy carries a very suggestive element; more fish remain in the sea than ever came out of it. It is not the habit of the evolutionary process to transform the whole of one state of living into another. Rather does nature pick some particularly happy development and allow it to expand in the place of and even at the expense of her earlier efforts. If man is to develop something new, the insistent question is, whether all humanity is going to develop or only a part of it? The biological analogy in favor of the latter would be overwhelming if man were an ordinary species, but it happens that at the moment, for the first time in history, he consists virtually of one society, and we have no precedent for the development of any new types, particularly of solitary types, from the middle of a single society; but what, of course, could develop from a society would be another society, at first simply a part of it, but afterwards differentiating itself more and more clearly.

If we consider only those alternatives that lead to development, leaving on one side the not impossible state in which mankind would be stabilized and live an oscillating existence for millennia, we have to consider, in the light of the present, the alternatives: whether mankind will progress as a whole or will divide definitely into a progressive and an unprogressive part. Over and over again in history there has occurred the raising of a particular class or a particular culture to a point at which there seemed a permanent gulf between it and other cultures or classes. Yet the gulf was not permanent; the particular aristocracy fell or its advantages spread themselves so widely that they became common stock. The cause for this is not obscure: first, the aristocrats differed only superficially from the many, and secondly they were not progressing themselves in such a way as to increase their distance and leave humanity behind. The present aristocracy of western culture, at the moment when it most clearly dominates the world, is being imitated rapidly and successfully in every eastern country. It is not on the lines of a cultural aristocracy or the formation of a class more able to lead the good life that the splitting of the human race is likely to occur; because such aristocracies are only reaching to a more complete humanity, and where they lead the race will follow. It is rather the aristocracy of scientific intelligence that may give rise to new developments. They have come down the earlier centuries, scattered singly or in small groups, but the mechanical revolution and its consequences have increased their number and at the same time their compactness. More and more, the world may be run by the scientific expert. The new nations, America, China and Russia, have begun to adapt to this idea consciously. Scientific bodies naturally are at first conceived of as advisory and they will probably never become anything else; but, with every advance in the direction of a more rational psychology, the power of advice will increase and that of force proportionately decrease. This development, coupled with the broadening of the idea of private interest to include, almost necessarily, some consideration of humanity, will tend to center real sovereignty in advisory bodies. The scientists would then have a dual function: to keep the world going as an efficient food and comfort machine, and to worry out the secrets of nature for themselves. It may well be that the dreams of *Dædalus* and the doom of *Icarus* may both be fulfilled. A happy prosperous humanity enjoying their bodies, exercising the arts, patronizing the religions, may be well content to leave the machine, by which their desires are satisfied, in other and more efficient hands. Psychological and physiological discoveries will give the ruling powers the means of directing the masses in harmless occupations and of maintain a perfect docility under the appearance of perfect freedom. But this cannot happen unless the ruling powers are scientists themselves. For a state in which the present rulers impose themselves in this way, the prospect of which so appalls Mr. Bertrand Russell, though possible, is essentially unstable and bound to lead to revolution, which would be brought about by the gradually increasing inefficiency of the rulers and the increasingly effective insurgence of the excluded intelligent. Even a scientific state could only maintain itself by perpetually increasing its power over the non-living and living environment. If it failed to do so, it would relapse into pedantry and become a perfectly ordinary aristocracy. In the earlier chapters I have given some idea of one way in which this scientific development could take place by the colonization of the universe and the mechanization of the human body. Once this process had started, particularly on the physiological side, there would be an effective bar between the altered and the non-altered humanity. The separation of the scientists and those who thought like them - a class of technicians and experts who would perhaps form ten per cent. or so of the world's population - from the rest of humanity, would save the struggle and difficulty which would be bound to ensue if there were any attempt to change the whole bulk of the population, and would, to a certain extent, lessen the hostility that these fundamental changes would necessarily produce. Mankind as a whole given peace, plenty and freedom, might well be content to let alone the fanatical but useful people who chose to distort their bodies or blow themselves into space; and if, at some time, the magnitude of the changes made them aware that something important and terrifying had happened, it would then be too late for them to do anything about it. Even if a wave to primitive obscurantism then swept the world clear of the heresy of science, science would already be on its way to the stars.

In tracing this development, however, we have neglected other weighty considerations. Up to the present the cumulative edifice of science has been erected by assistance as much from the practical world as from the learned, and scientists themselves have never formed an hereditary or even a closed caste. In two ways the progress of science depends upon non-scientific humanity. As experimentation becomes more complex, the need for the co-operation in it of technical elements from outside becomes greater and the modern laboratory tends increasingly to resemble the factory and to employ in its service increasing numbers of purely routine workers. If development is to follow, even in the earliest stages, on the lines I have indicated above, this necessity for economic and technical assistance will be multiplied many times. More important still, the complexities of scientific - and particularly of theoretical scientific - thought, calls for an ever greater number of first-class intelligences, and the modern development of science can hardly be disconnected from the political and economic changes which make it possible to recruit the personnel of science from wider and wider circles. For until we can know from the inspection of an infant or an ovum that it will develop into a genius, or else can from any infant produce one by a suitable education we shall have to rely on the diffusion of a general education in order to ensure that all capable minds are utilized.

This recruiting of science is the surest way of preventing a permanent human di-morphism from arising, because it reinforces what is probably the strongest factor involved, the emotional conservatism of the scientists themselves. The mere observation of scientists should be sufficient at the present to show that any fear of immediate di-morphism is unfounded. In every respect, save their work, they resemble their non-scientific brothers, and no one would be more shocked than they at the suggestion that they were raising up a new species and abandoning the bulk of mankind. For whether they are inventing submarines or depth charges, they feel they are serving humanity.

The consciousness of solidarity - and even more, the unconscious emotional identification with the group - is a terrific force binding humanity together, and so long as individual scientists have it, di-morphism would be impossible.

But the scientists are not masters of the destiny of science; the changes they bring about may, without their knowing it, force them into positions which they would never have chosen. Their curiosity and its effects may be stronger than their humanity.

These two obstacles to the separation of the scientists, though weighty, are of the kind that would lose force with time, while those favoring their separation tend to increase. The technical importance of the scientist is bound to give him the independent administration of large funds and end the mendicant state in which he exists at present. Scientific corporations might well become almost independent states and be enabled to undertake their largest experiments without consulting the outside world - a world which would be less and less able to judge what the experiments were about. It is very probable that before the real independence of science could make itself felt, the organization of the world would have to pass through its present semi-capitalistic stage to complete proletarian dictatorship, because it is unlikely that a scientific corporation would, in an ordinary capitalistic state, be allowed to be so wealthy and powerful. In a Soviet state (not the state of the present, but one freed from the danger of capitalist attack), the scientific intuitions would in fact gradually become the government, and a further stage of the Marxian hierarchy of domination would be reached. Scientists in such a stage would tend very naturally to identify themselves with the progress of science itself than with that of a class, a nation or a humanity outside science, while the rest of the population would, by the diffusion of an education in which the highest values lay in a scientific rather than in a moral or a political direction, be much less likely to oppose effectively the development of science. Thus the balance which is now against the splitting of mankind might well turn, almost imperceptibly, in the opposite direction. The whole question is one largely of numbers, and would become entirely so as soon as the quantity and quality of population were controlled by authority. From one point of view the scientists would emerge as a new species and leave humanity behind; from another, humanity - the humanity that counts - might seem to change *en bloc,* leaving behind in a relatively primitive state those too stupid or too stubborn to change. The latter view suggests another biological analogy: there may not be room for both types in the same world and the old mechanism of extinction will come into play. The better organized beings will be obliged in self-defense to reduce the numbers of the others, until they are no longer seriously inconvenienced by them. If, as we may well suppose, the colonization of space will have taken place or be taking place while these changes are occurring, it may offer a very convenient solution. Mankind - the old mankind - would be left in undisputed possession of the earth, to be regarded by the inhabitants of the celestial spheres with a curious reverence. The world might, in fact, be transformed into a human zoo, a zoo so intelligently managed that its inhabitants are not aware that they are there merely for the purposes of observation and experiment.

That prospect should please both sides: it should satisfy the scientists in their aspirations towards further knowledge and further experience, and the humanists in their looking for the good life on earth. But somehow it fails by the very virtue of its being a possible and probable solutions on the lines of our own knowledge. We do not really expect or want the probable; all, even the least religious, retain in their minds when they think of the future, an idea of the *deus ex machina,* of some transcendental, superhuman event which will, without their help, bring the universe to perfection or destruction. We want the future to be mysterious and full of supernatural power; and yet these very aspirations, so totally removed from the physical world, have built this material civilization and will go on building it into the future so long as there remains any relation between aspiration and action. But can we count on this? Or, rather, have we not here the criterion which will decide the direction of human development? We are on the point of being able to see the effects of our actions and their probable consequences in the future; we hold the future still timidly, but perceive it for the first time, as a function of our own action. Having seen it, are we to to turn away from something that offends the very nature of our earliest desires, or is the recognition of our new powers sufficient to change those desires into the service of the future which they will have to bring about?

Engels and Science

IF Engels had not been the constant companion in arms of Marx in the revolutionary struggles of the 19th century, there is no doubt that he would be remembered chiefly as one of the foremost scientist-philosophers of the century. It was an ironical tribute paid to the correctness of his views as to the relations between politics and ideology that he suffered complete neglect from the scientists of the Victorian age. But time now has taken its revenge, and Engels' contemporary views on 19th century science seem to us now in the 20th far more fresh and filled with understanding than those of the professional philosophers of science of his day, who for the most part are completely forgotten, while the few that linger on, such as Lange and Herbert Spencer, are only quoted as examples of the limitations of their times. It would, of course, be wrong to consider Engels' scientific achievement apart from his association with Marx. It was through Marx's influence, and by the methods of dialectical materialism they evolved together from Hegel's dialectic idealism, that he achieved the possibility of criticising and interpreting science in a manner which was not open to his predecessors.

Engels as a Scientist

It is often said by those anti-Marxists who never trouble to read the original writings that the scientific knowledge of Marx and Engels was superficial; that Engels, for instance, sought in later life for scientific justification for the dialectical laws that Marx had introduced into economics. This is a complete misreading of the facts. Engels' interest in and knowledge of science was deep and early. It ran through all his philosophical and political studies. In an essay as early as 1843 (quoted in the Marx-Engels, Selected Correspondence, p. 33), he shows a grasp of the fundamental connection between science and productivity that was to run through all his later work:—

. . . . yet there still remains a third factor—which never counts for anything with the economists, it is true—namely science, and the advance of science is as limitless and at least as rapid as that of population. How much of the progress of agriculture in this century is due to chemistry alone, and indeed to two men alone—Sir Humphry Davy and Justus Liebig? But science multiplies itself at least as much as population: population increases in relation to the number of the last generation; science advances in relation to the total amount of knowledge bequeathed to it by the last generation, and therefore under the most ordinary conditions in geometrical progression too—and what is impossible for science?

Engels to the very end of his life not only made use of the science he had learnt at the University, but kept up with extraordinary keenness and understanding his interest in the scientific discoveries of his times. Far from being prejudiced by any preconceived theories, he was more open to accepting new ideas than were the professional scientists. In a letter to Marx in 1858, he shows himself prepared to accept beforehand the idea of transformation of species which Darwin was to publish in the next year (Marx-Engels, Correspondence, p. ll4). In one passage he almost hints at the idea of evolution, derived from the Hegelian idea of transformation of quantity into quality:—

So much is certain; comparative physiology gives one a withering contempt for the idealistic exaltation of man over the other animals. At every step one bumps up against the most complete uniformity of structure with the rest of the mammals, and in its main features this uniformity extends to all vertebrates and even—less clearly—to insects, crustaceans, earthworms, etc. The Hegelian business of the qualitative leap in the quantitative series is also very fine here.

A few months later, when Darwin's "Origin of Species" appeared, Engels and Marx together acclaim it as putting an end to teleology in the natural sciences. Already Engels on December 12, 1859, exactly four weeks after the publication of the first edition, writes to Marx: "Darwin, whom I am just now reading, is splendid," and Marx writes in reply: "Although it is developed in the crude English style, this is the book which contains the basis in natural history for our point of view."[1]

If we contrast this attitude to that of the official philosopher of science and physicist, Whewell, a great derider of Hegel, who was at the same time urging that Darwin's book be not accepted by Trinity College Library, we can measure the greater breadth and penetration which their philosophical outlook had given to Marx and Engels. It was the same with all the significant ideas which science was developing. The great physical and chemical advances of the century, particularly the conservation of energy and the development of organic chemistry, were also recognised and carefully studied by Marx and Engels. In his approach to science, Engels cannot be said to have been an amateur. In Manchester, where he spent most of his life, there was a very lively scientific life with which he freely mixed, and, in particular, he had as his intimate friend Karl Schorlemmer, the first Communist Fellow of the Royal Society, and one of the most distinguished chemists of his time.

The width of Engels' scientific knowledge can be fully appreciated only from a study of his great unfinished work, *Dialectic and Nature*. In it different sciences are treated comprehensively and critically. It is easy to see from the authorities cited how close Engels was to contemporary developments in mathematical, physical, and biological sciences, to say nothing of sociology and economies. He even includes a short and amusing chapter on psychic science.

Engels on the History of Science

From the start Engels was able to unify his conceptions of science in such a way that he could naturally assimilate new developments as they appeared, and that without any of the wilder flights of such scientific philosophers as Haeckel or Herbert Spencer, but in an extremely sane and balanced way. The secret of this power lies in the materialist dialectic which he used in his analysis of the results of science. It was from Hegel that he learnt to appreciate, not things, but processes, and he always looked at the position which science had reached at any time in relation to its historical background. This is clearly seen in his essay on Feuerbach, where he traces the history of materialist philosophy in relation to the development of science and productive methods. For instance, he says:—

> But during this long period from Descartes to Hegel and from Hobbes to Feuerbach, the philosophers were by no means impelled, as they thought they were, solely by the force of pure reason. On the contrary. What really pushed them forward was the powerful and ever more rapidly onrushing progress of natural science and industry. Among the materialists this was plain on the surface, but the idealist systems also filled themselves more and more with a materialist content and attempted pantheistically to reconcile the antithesis between mind and matter. Thus, ultimately, the Hegelian system represents merely a materialism idealistically turned upside down in method and content. . . .

> The materialism of this last century was predominantly mechanical, because at that time, of all natural sciences, mechanics and indeed only the mechanics of solid bodies —celestial and terrestrial—in short, the mechanics of gravity, had come to any definite close. Chemistry at that time existed only in its infantile, phlogistic form. Biology

58

still lay in swaddling clothes; vegetable and animal organisms had been only roughly examined and were explained as the result of purely mechanical causes. As the animal was to Descartes, so was man a machine to the materialists of the eighteenth century. This exclusive application of the standards of mechanics to processes of a chemical and organic nature—in which processes, it is true, the laws of mechanics are also valid, but are pushed into the background by other and higher laws—constitutes a specific but at that time inevitable limitation of classical French materialism.

The second specific limitation of this materialism lay in its inability to comprehend the universe as a process—as matter developing in an historical process. This was in accordance with the level of the natural science of that time, and with the metaphysical, *i.e.*, anti-dialectical manner of philosophising connected with it. Nature, it was known, was in constant motion. But according to the ideas of that time, this motion turned eternally in a circle and therefore never moved from the spot; it produced the same results over and over again. (*Feuerbach*, pp. 36 and 37.)

As a historian of science Engels is particularly distinguished. He was the first to understand with Marx the close relation between the development of scientific theory and of productive methods. Much of what now passes for new in the interpretation of historical science is to be found in the pages, of *Dialectic and Nature*,[2] He notices, for instances, that the theory of heat did not develop from pure thought, but from a study of the economic working of steam engines, and comes to the conclusion: "Until now they have only boasted of what production owes to science, but science itself owes infinitely more to production."[3] In particular he shows how the metaphysical and statical attitude of the 18th century materialists based on Newton was broken down in favour of a view which reflects, though unconsciously, a dialectical progress: "The beginnings of revolutionary science faced a through and through conservative nature, in which everything is to-day as at the beginning of the world, and will be to the end of the world the same as it was at the beginning."[4] The breaches made in this outlook he indicates as, first Kant and Laplace's nebular hypothesis, second the development of geology and paleontology, third chemistry, which can synthesise organised substances and whose rules hold just as much for the processes of life, fourth the discovery of the conservation of energy, fifth Darwin's evolutionary theory, and sixth the synthesis of all the processes affecting life, animal ecology and distribution. The significance of the break is described as follows:—

It was not the scientists but the philosophers who made the first breach in this fossilised outlook. In 1755 appeared Kant's "General Natural History and Theory of the Heavens." The problem of the first impulse was here set aside. The earth and the whole solar system appeared as something become in the course of time. If, before the appearance of this thought, the overwhelming majority of scientists had not felt the fear expressed by Newton in his warning "Physics, Beware of Metaphysics!"[5]—then they would have drawn from this single discovery of genius by Kant such consequences as would have saved them infinite errors along circuitous paths, and an immense quantity of time and labour expended in a false direction. In Kant's discovery lay the germ of all further progress. If the earth was something which had become, then all its present geological, climatic and geographical condition had become also, its flora and fauna as well, and it must have a history not merely in space, but in time also. (Quoted by V. L. Komarov in *Marxism and Modern Thought*, p. 205. See also M.E.A., Vol. 2, p. 244.)

As a result of these movements of thought, Engels says:—

> The old teleology has gone to the devil, but now we have the knowledge that matter in its perpetual circulation moves according to laws that at certain stages—now here, now there—necessarily produce the thinking mind in organic existence. (M.E.A., Vol. 2, p. 175)

Engels' concept of nature was always as a whole and as a process. He escaped the specialisation which even in those days made it impossible for a physicist to understand biology or vice-versa, and he laid down a general outline of this process which can still be the basis for an appreciation of the results of scientific research.

He never had the opportunity to put down in one place his view of this universal process. The main outlines can be seen in *Anti-Dühring*, or even better in the shortened form of *Socialism, Utopian and Scientific*. But for its full appreciation in this country we shall have to wait until the publication in English of *Dialectic and Nature*. Throughout Engels wages war on metaphysical ways of thinking in science, with its fixed categories and its sharp distinctions between cause and effect, structure and behaviour, identity and difference, whole and part[6]. These are not so much invalid as valid only in small, defined regions. The success of the scientific method is best seen in such regions: "For everyday use, for scientific retail trade, the metaphysical categories still keep their value."[7] The dialectical approach to science has its value, on the contrary, in its comprehensiveness. The movements first seen by Hegel in the ideal world are, according to Marx and Engels, simply reflections of those in the objective world. Much of Engels' studies were devoted to exemplifying the Hegelian modes, particularly those of the transformation of quantity into quality, the interpenetration of opposites and the negation of negation, in the world of science.

In *Anti-Dühring* this is done in the shortest way. But the *Dialectic and Nature* contains far more examples.

The Transformation of Quantity into Quality

Philosophers still cavil at the use of the phrase "transformation of quantity into quality" on the grounds that it is not quantity that changes into quality, because quantity remains in the end. But the phrase is simply a shorthand way of referring to Hegel's law that purely quantitative changes turn into qualitative changes. It was in this form that Marx understood it, as shown explicitly in his letter to Engels (Letter 97). The examples which Engels gives, the case of ice turning into water, or water into steam, and that of the change of physical quality of a chemical substance with the number of atoms that are comprised in it, should have shown sufficiently clearly what this concept meant. With remarkable insight Engels says —

> The so-called constants of physics are for the most part nothing but designations of the nodal points where quantitative addition or withdrawal of motion calls forth a qualitative change in the state of the body in question. (M.E.A., Vol. 2, p. 288.)

We are only now beginning to appreciate the essential justice of these remarks and the significance of such nodal points. The whole theory of quanta depends, like the theory of acoustic vibrations with which it has formal relations, on the distribution of nodes which mark out two qualitatively and quantatively different states of vibration.

The problem of qualities had always raised the greatest difficulties to the philosophers and furnished, as it still furnishes, a reason for invoking outside forces. From any logical materialist standpoint it is necessary to recognise that a new quality of a system is something not in any sense added to the system, but produced simply by a continuous change in its already existing components. To make this meaning perfectly clear, Engels cites as his final authority Napoleon.

In conclusion we shall call one more witness for the transformation of quantity into quality, namely—Napoleon. He makes the following reference to the fights between the French cavalry, who were bad riders but disciplined, and the Mamelukes, who were undoubtedly the best horsemen of their time for single combat, but lacked discipline: "Two Mamelukes were undoubtedly more than a match for three Frenchmen; 300 Frenchmen could generally beat 300 Mamelukes, and 1,000 Frenchmen invariably defeated 1,000 Mamelukes." (*Anti-Dühring*, p. 146.)

Engels found many examples in science of this transformation. Of these I can only quote one, that of Mendeleyeff's Periodic Law, which was to prove in the future so rich in further examples of the transformation of quantity into quality.

Finally, Hegel's law holds not only for compound bodies, but for the chemical elements themselves. We know now that chemical properties of elements are a periodic function of their atomic weight and consequently their quality is determined by the quantity of their atomic weight (or, as we would now say, of their atomic number), and the proof of this has been made in a most striking way. . . . By the help of the—unknown— application of Hegel's law of the change of quantity into quality, Mendeleyeff has achieved a scientific feat which can well stand comparison with Leverrier's calculation of the orbit of the still unknown planet Neptune. . . . Perhaps those gentlemen who up till now have treated the transformation of quantity into quality as mysticism and incomprehensible transcendentalism will now explain that it is all perfectly self-evident, trivial, and platitudinous, that it has been long familiar to them and that we have nothing new to teach them. To have put forward for the first time a general law of nature and thought, in its most generally valid form, that will always remain as a historical achievement of the first order, and if these gentlemen for so many years have allowed quantity and quality to turn into each other without knowing what they were doing, they must console themselves with Molière's Monsieur Jourdain, who had all his life spoken prose unwittingly. (Engels' *Dialectic and Nature*, p. 289.)

Understood in this way, the concept of the transformation of quantity into quality can be, and is being, extremely valuable in scientific thought. We are learning more and more that specific qualitative properties of bodies depend on the *number* of certain of their internal components. If an atom can only link with one other atom, the result is a gas. If it can link with *two* or *three*, the result will be a solid of fibrous or platy character. If with *four*, a hard crystalline solid like diamond. If with *more than four*, a metal. Similarly the processes of freezing, boiling, vitrification, etc., depend on what are now known as "co-operative" phenomena. It takes a million or more molecules to make a substance which can be recognised as a solid or liquid: a smaller number leads to the qualitatively different colloid state.

The Interpenetration of Opposites

The concept of the interpenetration of opposites has not been given by Engels the same coherent treatment as that of the others. Yet it recurs nearly all the way through his scientific writings. It appears in two shapes, firstly, as the Hegelian idea that nothing can be defined apart from its opposite, that, so to speak, everything implies its opposite (here Engels approached very close to the modern ideas of relativity) but also more objectively that there exist no hard and fast lines in nature.

"Hard and fast lines" are incompatible with the theory of development. Even the border line between vertebrates and invertebrates is no longer unchanging. Every day the lines of demarcation between fish and amphibia, between birds and reptiles, tend more and more to vanish. Between the *Compsognatus* (a small dinosaur) and the *Archæopteryx* (a toothed bird of the same origin) only a few intermediary members are wanting, while toothed birds' beaks have been found in both hemispheres. (Quoted by V. L. Komarov in *Marxism and Modern Thought*, p. 199. See also M.E.A., Vol. 2, p. 189).

In physics Engels exemplified this principle by the example of magnetism, in which each N. Pole implies a S. Pole or vice-versa, or more generally in the balance between attraction and repulsion. Here, Engels' treatment is surprisingly modern. He understands forces not as mystical entities, but to be known only by the movements produced by them. This is characteristic of the modern tendency of turning mechanics into kinematics. In Engels' analysis attraction is simply the reflection of the coming together of bodies, as repulsion is of their separation. Thus heat in the kinetic theory of gases acts as a repulsive force.

The Negation of the Negation

It is the same with the principle of the negation of the negation, which Engels illustrates with the famous examples of the barley seed negating itself into a plant and the plant further negating itself into many seeds, as well as the mathematical examples of the product of negative quantities and the differential calculus. These are the kind of statements that until recently made dialectical materialism seem quite unacceptable, indeed incomprehensible to scientists trained along official lines. Negation has always seemed to them something only applicable to human statements, but this is just a defect of language. If we had a word to describe how something in the course of its own inner development can produce something else different and in some sense opposite to it, and which comes in time to replace it entirely, that word would take the place of negation. Negation in this sense is not a symmetrical operation; the negation of negation does not reproduce the original, but something now

65

unlike both. As long as we deal in mere words, however, such statements can convey very little. It is in concrete examples that the significance of the negation of the negation can effectively be grasped. And if Hegel's and Engels' works had been treated on their merits instead of as something to be attacked in every possible way, the sense of their use of "negation of negation" would have been clearly apparent. But this, of course, would also have meant the recognition of the necessity of revolution, and that was far too uncomfortable to be accepted.

Just as the transformation of quantity to quality, so the principle of the negation of negation finds many examples in modern science. In almost every physical process in nature, there is a tendency for the process itself to create an opposition which ultimately brings it to a stop, which in turn results in the disappearance of the antagonistic process and the re-establishment of the original one. Take, for example, the case of the building up of mountain ranges due to strain in the earth's crust. This results in increased weathering which destroys the mountain range and accumulates sediments which lead to further crust strains, leading to further mountain building, etc. Modern physics is full of dialectical contradictions of this type—wave and particle, matter and energy—and even in Freudian psychology the provisional analyses of the mechanism of instinct and its repression are stated in a dialectical form. The whole of modern science is unconsciously affording more and more examples of the aspect of phenomena that can only be consciously grasped through dialectical materialism.

The Dialectical Process of Nature as a Whole

But Engels did not confine himself to scientific illustrations of the validity of his philosophical position. His main task was a constructive one, and he gives in several places both in his Letters, in the *Anti-Dühring*, and the essay on Feuerbach, his general view of the dialectical process of nature taken as a whole. (See particularly Letter 232 and Chapters 5 to 8 of *Anti-Dühring*.)

Dialectic and Nature was intended to give such a complete conception, but it was never finished and contains as it stands a number of more or less filled-in sketches of such conceptions.[8] In the omitted fragment from Feuerbach (p. 76 of the English edition) he recapitulates the chief points in which the science of his time had served to lay the basis of a comprehensible materialistic view of the development of the universe. In this he lays stress on three discoveries of decisive importance:

The first was the proof of the transformation of energy obtained from the discovery of the mechanical equivalent of heat (by Robert Mayer, Joule and Colding). All the innumerable operative causes in nature, which until then had led a mysterious inexplicable existence as so-called "forces"—mechanical force, heat, radiation (light and radiant heat), electricity, magnetism, the force of chemical combination and dissociation—are now proved to be special forms, modes of existence of one and the same energy, *i.e.*, motion. . . . The unity of all motion in nature is no longer a philosophical assertion but a fact of natural science.

The second—chronologically earlier-discovery was that of the organic cell by Schwann and Schleiden—of the cell as the unit, out of the multiplication and differentiation of which all organisms, except the very lowest, arise and develop. With this discovery, the investigation of the organic, living products of nature—comparative anatomy and physiology, as well as embryology—was for the first time put upon a firm foundation. The mystery was removed from the origin, growth and structure of organisms. The hitherto incomprehensible miracle resolved itself into a process taking place according to a law essentially identical for all multi-cellular organisms.

But an essential gap still remained. If all multi-cellular organisms—plants as well as animals, including man—grow from a single cell according to the law of cell-division, whence, then, comes the infinite variety of these organisms? This question was answered by the third great discovery, the theory of evolution, which was first presented in connected form and substantiated by Darwin. . . .

With these three great discoveries, the main processes of nature are explained and traced back to natural causes. Only one thing remains to be done here: to explain the origin of life from inorganic nature. At the present stage of science, that means nothing else than the preparation of albuminous bodies from inorganic materials. Chemistry is approaching ever closer to this task. It is still a long way from it. But when we reflect that it was only in 1828 that the first organic body, urea, was prepared by Wöhler from inorganic materials and that innumerable so-called compounds are now artificially prepared without any organic substances, we shall not be inclined to bid chemistry halt before the production of albumen. Up to now, chemistry has been able to prepare any organic substance, the composition of which is accurately known. As soon as the composition of albuminous bodies shall have become known, it will be possible to proceed to the production of live albumen. But that chemistry should achieve overnight what nature herself even under very favourable circumstances could succeed in doing on a few planets after millions of years—would be to demand a miracle.

The materialist conception of nature, therefore, stands to-day on very different and firmer foundations than in the last century.

This quotation shows amply that not only had Engels a complete grasp of the essential stages of development up to the human level, but that he also saw very clearly the gaps in the explanation. The gaps are, first of all, the origin of the stellar universe as we know it, including the solar system and the earth, the origin of life on the earth, the origin of the human race, and the origin of civilisation. Each one of these questions was treated by Engels, and to each one he had valuable contributions to make.

The Origin of the Universe

Once dialectical materialism is understood, the logical absurdity of all creationist theories of the universe become apparent. It is not that dialectical materialism provides an alternate theory, but it shows that you cannot treat the Universe in the same way that you treat any part of it, as something acted on from outside. Whatever moves the Universe must be the Universe. In so far as it develops it is self-creating. In particular, it shows the childishness of assuming a personal Creator whether with the honest anthropomorphism of early tribal peoples or the reactionary idealism of the mathematician Godmakers of the present day. As Engels wrote: "Gott = Nescio, 'aber ignorantia non est argumentum' (Spinoza)."[9] At the same time he saw very clearly that there were social and political reasons for maintaining such beliefs, and of emphasising the helplessness of man before the existing state of nature and, by implication, the existing social and political order.

As to the origin of the universe, Engels put forward no new theory, but implied that the key to its discovery would lie in the study of the nature of matter and movement. Engels was from the beginning attracted to the nebular hypothesis, and enthusiastically took up the observations of spiral nubulæ of which our galaxy is only one example.

The Origin of Life

As the last quotation shows, Engels believed, at a time when that belief was far less plausible than it is now, in the chemical origin of life as a definite period in the earth's development. Short of a special creation of life, which had already become scientifically suspect by the middle of the 19th century, the only alternative theory was that life had always existed. This theory, upheld with the authority of Liebig and Helmholz,[10] Engels energetically combated. "Why should not," asked Liebig, "organised life be as old, as eternal, as matter itself? Why should it not be as easy to imagine this as the eternity of carbon, and its compounds?" To this Engels answered:

> (*a*) Is carbon simple? If it is not, it is as such not eternal. (*b*) Carbon compounds are eternal only in the sense that under such and such conditions of mixture, temperature, pressure, etc., they can be reproduced. However, only the simplest carbon compounds, for example CO_2 and CH_4, can be eternal because they can be at all times and more or less in all places, produced and decomposed into their elements. (M.E.A. Vol. 2, p. 180.)

He argues that with these exceptions the conditions for the production of carbon compounds will not exist except on the earth in living beings or in the laboratory, and that though their eternal existence is thinkable, this merely shows that anything that is thought need not necessarily exist. Far stronger is the argument against the eternity of albumen, which can exist only under the very narrow limits of temperature and moisture of the earth.

> The atmospheres of astronomical bodies, particularly of nebula, were originally white hot—no place for albumen —so that space must be the big reservoir, a reservoir lacking air and nourishment and at a temperature which no albuminous body can possibly exist. . . . What Helmnholz says of the unsuccessfulness of experiment in making life is just childishness. Life is the mode of existence of albuminous substances its intrinsic impetus comes from the continuous exchange of matter with the medium surrounding it, and with the ceasing of this exchange life itself ceases, and the albumen breaks up. (M.E.A., Vol, 2, p. 181.)

Time has not diminished the soundness of Engels—conclusions. We are still far from having analysed, much less synthesised, albuminous substances (for by that Engels did not mean protein in its modern sense as a pure crystalline chemical substance, but the complex of chemicals that underlie protoplasm —proteins, sugars, salts, etc. Nevertheless, through combination of modern biochemical knowledge with astrophysical and geological considerations about the early atmosphere of the planet, we can make a plausible picture of the origin of life by purely chemical means, and no other hypothesis for its origin can be put forward which will stand the slightest rational examination.

The Origin of Human Society

The next gap which Engels recognised was that in the development of human society from the animal stage, but it was not sufficient on this point to see and appreciate at their true value the results of scientific workers: here Engels was a scientist on his own account. The prevalent popular view in the 19th century was still that of the special creation of man. The materialists, led by Darwin, Huxley and Haeckel, maintained that man was only a superior ape distinguished by a larger brain. This brain which gave man his peculiar character was just such a product of evolution as a bat's wings or an elephant's trunk. Engels and Marx saw this crude explanation was hardly better than the theological one. They saw, long before anthropologists had taken up the question, that there was something qualitatively different about man which distinguished him from other animals, and that this was not an immortal soul, but the fact that man does not exist apart from society, and is in fact a product of the society which he has himself produced. Men, by entering into productive relations with each other, by the first exchange of food, and by the transmission of social characters through the family, became qualitatively different from other animals. These subjects were dealt with by Engels in an essay on "Work as the factor making for the transformation of Apes into Men," and in his most brilliant scientific work, *The History of the Family.*

V. L. Komarov, in his article on "Marx and Engels on Biology"[11] discusses at length this very point. The first stages, the development of man as a tool-using animal and as an animal capable of communicating with his fellows, can only be looked at from the biological point of view. It is at the same time the anatomical possibility inherent in a tree ape that has become a ground ape that make the use of instruments possible, and the use of instruments make the development of the human hand into its present form possible, without which it must have developed either hoofs or paws:

So the hand is not only an organ of labour; it is also its product. . . But the hand was not something self-sufficient: it was only one of the members of a complete and unusually complex organism, and what assisted the hand also assisted the whole body which the hand served, and assisted it in a double respect. (M.E.A., Vol. 2, p. 201.)

But at the same time, the development of manual skill inter-acted with the formation of primitive society.

The development of labour necessarily assisted the closer drawing together of the members of the society since because of it instances of mutual support and of common action became more frequent and the advantage of this mutual activity became clear to each separate member. To put it shortly, men when formed, reached the point when they felt the need of saying something to one another. The need created the organ. The undeveloped tongue of the ape was slowly but steadily changed by means of gradually increased modulations and the organs of the mouth gradually learned to pronounce one distinct sound after another. (V. L. Komarov, *Marxism and Modern Thought*, p. 201).

The Origin of the Family

In *The History of the Family* Engels takes up the story again at a later stage. It is here that the full value of Engels as a scientist can be appreciated. Long before its recognition by the official anthropologists, he appreciated the significance of the matrilinear family group or clan that travellers and missionaries were showing to exist among all primitive peoples. With his wide historical learning he linked these facts with the history of early Greece and Rome, and showed first of all what an admirable economic unit the matrilineal family was at a certain primitive stage of production, and secondly how it broke down first to the patriarchal family, and finally to the modern small family, under the influence of the development of property, itself due to better methods of production. All the more recent work of anthropologists and historians has only served to confirm Engels' original ideas. The transformation from the matrilinear family to the present form has been traced also in China and can be seen in actual course of operation in all primitive societies in contact with European civilisation, as Malinowski in particular has shown in great detail. Engels' anthropological studies were not merely academic exercises: they were closely related to the great task that he shared with Marx, the transformation of capitalist into socialist society. In recognising the relatively happy, courteous, and upright life of savages compared to their civilised descendants, he conceives the task of socialism as that of the return, again through the negation of the negation, to the nobility of the savage, without the sacrifice of the material powers which capitalist development had presented to mankind. His historical studies, particularly *The History of the Mark*, all led to the effecting of this transformation. He realised its difficulty (Letter 227):—

History is about the most cruel of all goddesses, and she leads her triumphal car over heaps of corpses, not only in war, but also in "peaceful" economic development. And

73

we men and women are unfortunately so stupid that we never can pluck up courage to a real progress unless urged to it by sufferings that seem almost out of proportion.

Engels' Work and the Development of Science

What is the relation of Engels' work to the enormous development of science that has gone on since his time? What has already been said should be sufficient to show that this has only confirmed the value of his methods of approach and suggested their further application. For part of the intervening period this has been done by Lenin in *Materialism and Empirio-Criticism*, or by the writings of Plekhanov and Bukharin. At the moment this work is being carried forward both theoretically and practically by the younger Soviet scientists.[12]

There is no doubt that Engels would have recognised and welcomed the main advances in the scientific field which have occurred since his time. He would have recognised that four significant steps have been taken. The Relativity theory has finally dethroned the mechanical materialism of the Newtonian school, but only in its mechanical and not its materialist aspects. Engels, who welcomed the principle of the conversion of one form of energy into another, would equally have welcomed the principle of the transformation of matter into energy. Motion as the mode of existence of matter would here acquire its final proof. The second great advance, the whole modern atomic and quantum theory, would also appear to him as a vindication of dialectical materialism. The diverse qualities of the natural elements now find their explanation simply in the number of electrons which compose them. Even more clearly than in organic chemistry, the transformation of quantity into quality is exemplified. The great advances in bio-chemistry which show the phenomena of living animals and plants as functions of the properties of the chemical molecules which make them up is a direct exemplification of what Engels had written about the chemical basis of life. Finally, the discovery of the mechanism of

74

inheritance through the chromosome theory (originally put forward by Mendel and now actually verifiable by microscopical observation) provides the material mode of transformation by which living animals develop and reproduce. These advances leave the main gaps in our knowledge still open, but we see more clearly than Engels could how they are likely to be filled. Nevertheless, Engels' work remains not only notable in its own time, but as valuable to us now in trying to keep the same all-embracing and historical approach to science that he possessed, and to use the methods he elaborated in pushing forward the solution of further problems.

After half a century of neglect, the methods of Engels and Marx are at last coming into their own in the scientific field. First, in the Soviet Union, but already also in England and France, the classics of dialectical materialism are being studied for the light they throw on present problems. In France in particular there have already appeared two notable contributions in *A la Lumière du Marxisme* (In the Light of Marxism) by a number of scientific writers and historians, and *Biologie et Marxisme* by Prenant. The crises of modern science appear in the first place as intellectual difficulties arising from new and apparently incompatible discoveries. The resolution of these crises, that is, the process of bringing them into harmony with the general movement of human thought and action, is a task for the Marxist scientists of to-day and to-morrow. The task is an endless one, and yet definite stages of advance can be established. We have through dialectical materialism a greater comprehension of whole processes, which before were only seen in their parts.

But it is not only in these general, almost philosophical, aspects of science that Engels' work is of value. In everyday work, those who take the trouble to follow Engels' hints find themselves more able to grasp the detailed connections of special investigations. The function of dialectical materialism is not to take the place of scientific method, but to supplement it by giving indications of directions in which hopeful solutions may be looked for. As Uranovsky says in *Marxism and Modern Thought*:

> The dialectic of nature is a method of the investigation and understanding of nature. This conception of nature is founded on the application of materialist dialectic to the data of science as they are obtained at each given historical moment. The dialectic of nature brings no artificial connections into nature and does not solve problems by substituting itself for the natural sciences. It helps in critically understanding and connecting facts already obtained, it points out the paths of further investigation and fearlessly poses uninvestigated problems. (p. 153.)

It is for the scientific method to judge whether these solutions are or are not true.

By showing how science has grown up as it were unconsciously in relation to these productive forces, it shows at the same time how this unconscious purpose, once grasped, can be consciously directed. This is what is happening in the U.S.S.R., and, once fully in action, it will be found that science has reached a new plane in its development.

But that stage will not come of itself; it will require intelligent collaboration on the part of the scientists themselves. In doing this they will make the memorial to Engels which is most in keeping with his spirit. For Engels was more than a scientist and a philosopher; he was a revolutionary. With him science acquired a new and positive meaning. As the last thesis on Feuerbach has it:

> "The philosophers have only interpreted the world in various ways. The point, however, is to change it."

76

Footnotes

1. Quoted by V. L. Komarov in Marxism and Modern Thought, p. 193. See also Marx Engels, Correspondence, Letter 49.

2. Marx and Engels Archives (German edition) Vol. 2, pp. 173, 194, et seq.

3. M.E.A., Vol. 2, p. 195.

4. M.E.A., Vol. 2, p. 175.

5. The use of the word metaphysical in Marxist literature is apt to cause confusion at first reading. The accepted popular use of the word is to connote assumptions which cannot be verified by concrete experience, generally, also somewhat vague and mystical assumptions. This is the sense in which it is used here and also the sense in which Marxism itself is said to be—quite wrongly—metaphysical. The Marxist use of the word, however, is more specialised. As can be seen from the quotations in this pamphlet, it is used only for a class of assumptions and categories that are abstract, fixed eternal and capable of absolute contradiction, such as the categories of Aristotelian logic or pre-relativistic physics. In contrast to these are the fluid dialectical categories.

6. M.E.A., Vol. 2, pp. 150 et seq.

7. M.E.A., Vol. 2, p. 189.

8. M.E.A., Vol, 2, pp. 134, 153, 216.

9. M.E.A., Vol. 2, p. 169. "God = I don't know, but ignorance is no argument."

10. M.E.A., Vol. 2, pp. 176 et seq.

11. Marxism and Modern Thought.

12. See for instance Science at the Cross-roads (Kniga 1931); and Science and Education in Soviet Russia, by A. Pinkevitch (Gollancz); and Marxism and Modern Thought, already quoted.

Dialectical Materialism and Modern Science

ONE of the questions on which clarity of thinking is now most necessary is that of the relation between the methods of science and of Marxist philosophy. Although much has already been written on the subject, yet there is still an enormous amount of confusion and contradictory statement. It is widely felt outside Marxist circles that, whatever the economic and political value of Marxist teaching, its incursion into the field of science is unwarranted. This is most strongly felt in relation to natural science, but it extends also to the social sciences in so far as these tend to imitate in their techniques the methods of natural science. Marxism is taken to be just another philosophic intrusion, adding nothing of importance and essentially superfluous in a region where the existing development of scientific method gives all the analysis which is necessary for the understanding of nature. Such an attitude, which has indeed been held by many who call themselves Marxists, implies at best a superficial view of Marxism and a lack of appreciation of its comprehensive nature. Much of this misunderstanding arises, particularly among those who have been trained in the English empirical tradition, from the fact that Marxist philosophy arose in part from Hegel and still retains a Hegelian terminology. The new direction which Marx gave to Hegelian philosophy and the solid material basis which he established for it are neither understood nor appreciated by those who are frightened by the phrases of "the transformation of quantity into quality" or "the negation of the negation." Those writers, on the other hand, who have attempted to remove from dialectic materialism its particular terminology generally also succeeded in removing the specific contributions which it has made to the understanding of the process of the universe and reduce it to a merely generalized application of normal scientific method. Now Marxism is not scientific method, nor is it in any sense an alternative method; it is at the same time more comprehensive and more advanced. Both the method of science as hitherto understood and the content of scientific discovery can be incorporated in the Marxist scheme. They need, however, to be criticized and extended. Marxism is no substitute for science, but because of its wider scope it can see the limitations of exiting methods and indicate where in the past these have been used in fields in which they have no competence. Further, it serves to complete the picture given by science by introducing into it a number of concepts and methods of working which have been, for historical and technical reasons, up till now foreign to it--and lastly to show science that its social function is not only contemplative but active. This is not to be taken to mean that Marxism is not science or that it is something which could be added on to science; or to set up an antithesis between Marxism and science. Marxism transforms science and gives it greater scope and significance, but we are not concerned here so much with this transformed Marxist science as with science as it is today.

One of the special features of Marx's work, which at first sight would seem to be an indication of the impossibility of the claims here advanced, was that he derived his analysis of the universe from the study of the development of human society. Human society is intrinsically more complex than any other part of nature, not only because it contains in itself all its complexities and more, but because its changes are more rapid and less regular. It is no accident that the sciences purporting to deal with it were the last to develop and are still the most unformed. Now science has proceeded almost axiomatically on the ground that the complicated is to be understood in terms of the simple and not vice versa. In doing so, however, especially in establishing those regularities which we know as scientific laws, it has necessarily deprived itself of the possibility of examining the type of phenomena that are not regular, particularly the appearance of novel elements in the universe. Now the rate of appearance of

80

novelty is itself the function of the complexity of the phenomena. We have no reason to believe that the vibrations of electrons in an atom of hydrogen have been for the last 10^{10} years different from what they are observed to be now. The progress of science, beginning with physics and working upwards to biology, did rest on the tacit assumption which was that of Aristotle and Averroes, that everything in the universe had proceeded and did proceed by unvarying and eternal rules. Anything therefore which did not depend on such rules was ipso facto excluded from the realm of science. Human history, for instance, was considered, except by aberrant intellects like Vico, to be an art and not a science. Even the cosmic evolution of Laplace did not seriously shake this position, because in his scheme it took place only as the result of the rigid application of the eternal Newtonian laws of motion. It was this attitude in fact which prevented for many hundreds of years the acceptance of the intrinsically obvious theory of organic evolution. But the evolution of new forms in the living world still remained as it remains largely today, a matter of inference and not of direct observation. The bulk of biological work on evolution has been rather to establish its reality and map out its line of advance than to inquire as to why it occurs at all. It is in fact only in the phenomena of our own society that we are able to see the development of radically new things occurring under our eyes, and if we are to understand how new things are produced in the universe it can only be, in the first place through such a study.

The way in which thinkers have approached the problem of history has gone through very curious and significant changes. In early times history was considered first as a storehouse of nobility and tribal self-glorification, and then for its value in moral edification. The first theories of history were justifications of the ways of God with men. It gradually appeared to the rationalists of the eighteenth century that this was not good enough, that making Providence responsible for everything in fact explained nothing. But they were not able to put anything very satisfactory in its place. The degradation of mankind through the appearance of wealth, kings, and priests was only a repetition on another plane of the story of the Fall. The scientific historians of the nineteenth century preferred to have no theory of history at all, and it degenerated into a chronicling of events which ceased to have any justification except giving employment to its professors. This was not entirely mental laziness; it betrayed a half-conscious apprehension that if people inquired too closely into the forces of human development they might find things inimical to the existing order.

Being from the outset free from this fear, Marx was able to see more in history than a meaningless sequence of events or vague tendencies towards progress. It was clear to him that he was not dealing with a unitary movement towards some foredestined end, but with conflicts which were resulting in the creation of new forms. The initial difficulty however remained, that before anything adequate could be discovered about the laws of these movements the phenomena themselves had to be ordered and grouped. It was for this purpose that he used the philosophy of his youth, though in doing so he transformed the most essential parts of the Hegelian world concepts. Hegel had introduced a most valuable and convenient classification. He saw the world in a hierarchical order. In other words, he was aware that the progress from simplicity to complexity is not an undifferentiated increase but can be divided naturally into successive stages, each stage having a general mode of behavior of its own. Each element in the hierarchy includes all those below it and is included in all those above. But the Hegelian hierarchy, because it was one of pure thought, could have no true development in time. The different stages were eternal and instantaneous. Marx, by making his hierarchy material, made it at the same time dynamic and historical. Each higher stage had actually grown out of the lower stage, and the new qualities it possessed were a product of those of the lower stages and of their mode of coming together. Thus the classes of human society are not just stock assemblages of people occupying a certain level in a social ladder but are the product of a tribal organization destroyed and reformed by the development of economic relations which had arisen from the development of the tribal economy itself. The categories with which Marx dealt differ from those used in science in that they are incapable of complete isolation. They must always be considered in relation to their origin and to their future development.

Now as science itself has proceeded almost entirely by the method of isolation and precise definition of categories independent of time, the Marxist method of thinking has *appeared* loose and unscientific, or as most scientists would put it, metaphysical. Isolation in science however can only be achieved by a rigorous control of the circumstances of the experiment or application. Only when all the factors are known is scientific prediction, in the full sense, possible. Now it is quite clear that were new things are coming into the universe all the factors cannot be known, and therefore that the method of scientific isolation fails to deal with these new things. But from the human point of view it is as necessary to be able to deal with new things as with the regular order of nature. It is perfectly right to restrict the use of the scientific method as it exists to the latter, but it is wrong to imply that outside this regular order the human mind is helpless, that if something cannot be dealt with "scientifically" it cannot be dealt with rationally. The great contribution of Marxism is to extend the possibility of the understanding and control of phenomena to include those in which radically new things are happening. This can only be done, however, subject to certain necessary limitations. In the first place, the degree of prediction where new things are concerned can never be of the same order of exactitude as in the regular and isolated operations of science. Exact knowledge which has been looked on as an ideal is however not the only alternative to no knowledge at all. There are, of course, very large regions inside science itself where exact knowledge is impossible. The whole trend of modern physics has shown that it is hopeless to expect it in atomic phenomena. But there the difficulty is circumvented by relying on the exactness of the statistical knowledge of a large number of events, and abandoning any claim to prediction of particular events. The exact dates and locality of the critical changes, the wars and revolutions that affect human society, are also unpredictable, and as there is only one human society even statistical methods are not strictly applicable. Nevertheless, the instability of certain economic and political systems call be shown to be due to intrinsic factors, and their breakdown becomes, within a wide limit of years, inevitable.

There can be no question, even to those completely unaware of the methods by which these predictions are reached, that the Marxists have some way of analyzing the development of affairs that enables them to judge far in advance of "scientific" thinkers what the trend of social and economic development is to be. The uncritical acceptance of this, however, leads many into believing that Marxism is simply another Providential theology, that Marx had mapped the necessary lines of social and economic development which men willy nilly must follow. This is a complete misunderstanding: Marxist predictions are not the result of working out such a scheme of development. On the contrary, they emphasize the impossibility of doing this. What can be seen at any given moment is the composition of the economic and political forces of the time, their necessary struggle and the new conditions which will arise as a result of it. But beyond that we can only foresee a process which has not ended and will necessarily take on new and strictly unpredictable forms. Marxism is valuable as a method and a guide to action, not as a creed and a cosmogony.

The relevance of Marxism in the development of science is both theoretical and practical. It removes science from its imagined position of complete detachment and shows it as part, a critically important part, of economic and social development. The complete revolution of the history of science as the result of Marxist analysis, so brilliantly summarized in Professor Hogben's article in *Science & Society*,[1] is one of the first results of this new attitude. But for Marxism understanding is inseparable from action, and the appreciation of the social position of science leads at once in a socialist country, such as the U.S.S.R., to the organic connection of scientific research with the development of socialized industry and human culture. The organization of science in capitalist countries has gradually molded itself in the service of big business, but because the process is not understood or appreciated its service is poor and incredibly wasteful. In any case production for profit can never develop the full potentialities of science except for destructive purposes. The Marxist understanding of science puts it in practice at the service of the community and at the same time makes science itself part of the cultural heritage of the whole people and not of an artificially selected minority.

The direct application of Marxism to scientific research is still very ill understood. It is clear that the scientific method as explicitly taught, while valid in establishing connections between phenomena, offers in itself no way of arriving at those connections. This fact is conveniently slurred over in scientific literature. In every scientific paper the data are given, the arguments from the data to the conclusions, and the conclusions themselves. What is not given, in general, is how the investigator chose the problem and how he thought of deriving the conclusions, and when reasons are given they are very rarely those actually used in the research but rather the formalized version of what the procedure of an ideal rational man would be in the circumstances. The whole drive of scientific inquiry is left implicitly to be explained by the operations of genius or intuition. The scientist actually does think of the new things, and it is no one's business to inquire why he does. This is where dialectical materialism comes in. Its value is not merely critical, as is classical scientific method, but indicative. It points the way in which it may be useful to look for new solutions. It is able to do this because of its way of linking up different aspects of nature under its general categories. It is extremely difficult to give examples because of the complexity of all the processes of scientific discovery, but from my own experience I have found Marxist methods invaluable for arriving at new conceptions. In the theory of liquids, for instance, we have to deal with phenomena that are not resolvable into the reaction of a particle with a certain environing force field but are strictly collective phenomena in which we have to consider at the same time the behavior of every particle and their mutual relations. It will be possible, when some systematic mind has been able to work on the subject, to develop out of the Marxist analysis a number of common scientific modes with some indication of which should be invoked in different circumstances. Collective behavior will obviously be one of these, another will be what might be called nuclear phenomena where the beginning of anything from the crystal to a revolution depends on a local assemblage of peculiarly favorable circumstances which alone enable it to get through the critical stages before which it is too small to grow.

Marxism has still another connection with science, that of criticizing its philosophic bases and the implications which seem to arise from the internal development of science itself. Marx, Engels and Lenin were all deeply concerned with this question, and for Marxist scientists of our time, though they have been distracted by the immediate needs of the economic situation in the Soviet Union or by the political situation outside it, it still remains a task of the greatest importance. On the fringe of science, and to the layman indistinguishable from it, are the pronouncements which the scientist makes on questions which are felt to be of vital human interest--those of the origin and fate of the universe, the nature of life, the character and behavior of the human mind and of society. In nearly every case the exact analysis of the statements reveals them as having little factual content, and in most cases they represent the dressing up of old traditional metaphysical ideas in the language, though not in the sense, of modern discovery. Such conceptions can be ruthlessly exposed and criticized from the Marxist point of view, because they represent entirely illegitimate use of science. One particular method of argument which is extremely common nowadays is that which establishes the existence of the supernatural from our ignorance of the natural. It is just in those spheres of science where the least exact knowledge exists that the strongest attempts are made to use science to bolster up ancient superstitions. Fortunately, it is just in such places that Marxist methods of attack are most valid, because they are all places where new things are being produced and where isolation so common in scientific research most palpably breaks down. These were all questions to which Marx and Engels devoted particular attention, and the way in which they were able to anticipate the trends of discovery in these fields is a striking indication of the value of the dialectical method. The modern Marxists have before them far vaster and more complex problems than had the pioneers. It seems probable that in the face of them modern science may well reach an impasse comparable with that which overcame the science of classical times. It is for the Marxists to find new methods of thought, of scientific organization and material technique which will prevent this happening.

The four critical points of the modern world view of science are the basic concepts of physics, which are now indissolubly bound up with the origin of the universe, the origin of life, the origin of human society and the fate of human civilization. In the first field it is more than ever clear that physics and astronomy are at present in an impasse. The contradictions between theory and observation in the field of cosmic rays, the expanding universe and the relation between fundamental physical units can no longer be obscured. Such contradictions are of course of enormous value to science, because out of the struggle to solve them will emerge some new and further-reaching generalizations, but until this happens no inferences can logically be made as to such ultimate questions; and even when it does, it can only be raising further and hitherto unglimpsed problems. Nevertheless, it is just this ignorance which is being used by the mystical physicists and astronomers to build a new creation myth. Just because the physicist cannot say, because the laws are not sufficiently well known, how the universe developed into its present state, they infer that it must have been created, as if this explanation did not raise enormously greater difficulties. From the Marxist point of view the problem of the origin of the universe in any ultimate sense is a pointless one. At any given stage the necessity of development of certain forms--stars, galaxies--may be derivable from the internal contradictions of some previous state, but there is no necessity to postulate either the eternal existence of a universe essentially like ours or a single ultimately primitive state. Indefinite regression of opposition and synthesis remains before us to explore.

The result of the progress of science in the last few centuries has been to progressively reduce the amount of work which the gods or God have had to do, but even yet the logical conclusion is not drawn. Evolution removed the necessity for special creation, but it is still considered that the Creator must have intervened to start the process off. Life appears as so qualitatively different from dead matter as to require some special act in its production. This problem again seems unreal to the Marxist, not that he denies the qualitative difference but that he sees in its origin just another example of that transformation of quantity to quality that is the characteristic of the appearance of new things. Life is sharply marked off from non-life, largely because its own operations effectively destroy the possibility of its continual recreation. In the primitive, lifeless world chemical substances were accumulated of the kind that cannot accumulate now because they would be consumed by the very life which their coming together in special circumstances brought into being. The practical scientists of today are learning to manipulate life as a whole and in parts very much as their predecessors of a hundred years ago were manipulating chemical substances. Life has ceased to be a mystery and has become a utility.

There yet remain the problems of man. The nineteenth century evolutionists certainly went too far in their demonstration that man was but a modified ape. The theologians were right in feeling that in this explanation something had been left out, but the soul which they postulated was again one of these mystical explanations which explain nothing. What Marx and Engels saw was the real qualitative difference between man and the animals was not the mere possession of a larger brain but the organization of human society; that human society was a category definitely different and higher than the animal species; that man in society represented a qualitatively new thing in the universe. The whole of modern anthropological and psychological research reinforces this conclusion: man is man-made, individually in the family, and socially through tradition and history, molded by his economic necessities and the means he has found to satisfy them.

Footnotes

1. "Our Social Heritage" by Lancelot Hogben. Science & Society, i, no. 2, 137-51.

Psycho-Analysis and Marxism

In the decade after the war Freud's theories dominated the narrow circles of British intellectuals. His psycho-analysis was accepted warmly for many reasons. It was new and exciting, it was shocking, it debunked religion and morals, it promised an internal liberation from all restraints. Nevertheless, it was essentially a creed of escape into an inner world of complexes and repressions and away from social and economic realities.

In recent years the Freudian wave has begun to recede. The effects of the world economic crisis of capitalism, and of the close menace of fascism and war, startled the intellectual strata into awareness of the objective world, and aroused a new wide interest in marxism, which the polite educated world had hitherto conspired to ignore and now approached like a new discovery.

It was inevitable, however, from this ill-digested process of thought that the demand should arise to "reconcile" Freud and Marx. The present book is an expression of this stage of extremely immature and uninformed confusion.

Great as the spread of marxism has been in the past few years it has by no means gone far or deep enough. Marxism is disturbing to existing habits; and it is only to be expected that once plain rejection and suppression are found no longer possible, attempts are made to water Marx down—to reconcile his ideas with existing fashionable modes of thought. Thirty years ago in Russia Machian positivism, so devastatingly castigated by Lenin in *Materialism and Empirio-Criticism*, was the fashionable intellectual creed; to-day we have the Freudian psychology. It is nevertheless a sad comment on the backward state of marxist knowledge in this country that such a book as Freud and Marx could be written at all, and still more serious that it should be so warmly recommended by such a well-known marxist writer as John Strachey.

91

* * * *

Mr. Osborn writes as if Freud and Marx had never been considered in juxtaposition before. He has never read, or shows no sign of having read, any of that voluminous amount of discussion on the subject already published in the Soviet Union and elsewhere. Worse still, however, almost every line of the book reveals a purely superficial understanding of marxism and a complete failure to grasp its essential principles.

He has, in effect, given a brief and far from adequate account of Freudian theory, which he accepts quite uncritically. There follows an interpretation, in hybrid marxist terms, of Primitive society, Historical Materialism and Dialectical Materialism. The general thesis is that Freud and Marx are to be reconciled in a dialectical way as two opposites, one representing the psychological and the other the material understanding of humanity. Out of the fusion of these is to come a superior understanding, some applications of which are given us in the last chapter.

Put in its strongest form, the argument is that psycho-analysis gives us a scientific interpretation of human motive which was not available to Marx and Engels. Had they been alive to-day they would—so it is argued—have accepted it as they accepted Darwinism. As they are not alive, the duty devolves on us to demonstrate by what processes they would have reached the stage of acceptance. There is, however, all the difference in the world between the methods of Marx and Mr. Osborn. Marx, while welcoming Darwin's scientific results, was never for a moment taken in by his philosophy. For his part, Mr. Osborn accepts Freud's philosophy without apparently realising how completely the work of Marx and Engels has already made it untenable.

The issue is a fundamental one. Freudism can—to an even less extent than Darwinism—not be treated as an experimental science, to be incorporated like Physics or Chemistry in the marxist interpretation of the objective world. Still less can Freudism be regarded as a dialectic complement to Marxism. For all its apparent materialism it is in effect just one more form of subjective philosophy and must be understood and rejected as such. This is not to deny the greatness of Freud's own work in clinical psychology or the importance of some of the relationships in human behaviour which it has brought to light. These, though they cannot by themselves be said to constitute a science of psychology, are a contribution to science. In order to make use of them we need to separate them carefully and critically from the almost mythical and continually changing theories that are involved in their presentation.

* * * *

The implication of this book is that marxism is deficient in psychological interpretation and that this deficiency can be met by psycho-analysis. In actual fact it is through Marx, and not through Freud, that we can begin to understand the significance and the possibilities of psychology. Marx does not start, as Freud does, with the idea of an essentially unalterable human psychology from which sociology can be derived. On the contrary he makes humanity for the first time comprehensible as the new quality which arises from social aggregation. We must understand society before we can understand man.

Human nature is not constant; it can be and is being moulded by society. Freud is incorrect when he produces, from his study of the psychology of the bourgeois family a generalisation to fit the whole human race, reaching as far back as the hypothetical primal horde with its jealous and terrible father. To accept the Freudian analysis is to accept by implication a completely non-dialectical view of psychology which must destroy the whole basis of marxist analysis. While ignoring the development of the world process, the rise of classes and the struggles between them in which human nature is formed and transformed, the Freudian interaction seeks to set up the individual, and the bourgeois individual at that, as the centre and measure of all things.

It is clear that Mr. Osborn has not only failed to notice this basic incompatability, but throughout his book he even tries to adduce a detailed parallelism between Freudian and marxist ideas. Many of these attempts are glaring distortions, as for instance when he seeks to establish the similarity of Freud's and Engels' views on the origin of the family, whereas in fact these views are almost diametrically opposite. Even more glaring is the attempt of Strachey in the preface to equate the "false consciousness" of Engels with the Freudian unconscious.

* * * *

Though more easily recognisable, the political aberrations that are contained in the second part of the book are merely natural consequences of this initially defective theoretical approach. Nevertheless, some are sufficiently extreme to warrant special mention. The author identifies the influence of traditional, and consequently counter-revolutionary, forces with the Freudian super ego, thence drawing the conclusion that it is necessary to weaken the super ego and substitute ego-morality for super ego-morality. This is apparently considered to represent or amplify the marxist idea of class-consciousness. There could not be a more gross distortion.

The essence of marxism is not that it substitutes one psychological attitude for another but that it provides an objective and scientific picture of the processes of social change—of the inevitable breakdown of capitalism and the role of the working class in bringing it about. As a result of this disintegration and through active participation in the political and economic struggles of the workers, old loyalties give place to new, but the new loyalties are on an altogether different plane of consciousness.

Related to this misconception of marxism is the idea that

as the material necessity for socialism is now overwhelming, all the resistance to the process of socialisation must therefore be psychological, and that psychology should consequently now play a decisive part in the struggle. In the first place this analysis of the current situation is wide of the facts.

The rising wave of working-class activity in all countries of the world to-day springs, like other previous waves, from a keen awareness of the economic situation and its effect on the workers. But there is to-day a far wider and deeper consciousness of the instability of capitalism than ever before. What dams up the wave is not bad psychology but the tardy development of the workers' political organisation, disunity, and the widespread prevalence of the ideas of social democracy and class collaboration. To suggest an appeal to psychology at the present time is to attempt to graft on to the tactics of Marxism an entirely subjective factor.

This is the logical deduction from the false antithesis of subjective and objective which underlies most of the book. To the marxist the subjective world is not opposed to, but part of, the objective world, and this is recognised in practice by the inclusion of psychology in revolutionary tactics. The idea of psychology as an independent dynamic element in politics leads straight to the "change of heart" school—the pacifists and liberal apologists of capitalism.

* * * *

Actually the author is led further than this, for he advocates as the chief practical application of his theories a Communist leadership which in everything but the words constitutes the Führer principle of the Fascists. The "leader" is to be made into a father figure in whom his followers are to have infinite confidence and such an attitude would completely justify the reactionary propaganda of those who reject "dictatorship of the Right and of the Left" because they see no difference between

Fascism and Communism.

In actuality the principles of leadership under Communism on one hand and Fascism on the other, are fundamentally opposed. The whole psychological apparatus of the Fascist "leader" is designed to deceive his followers and to distract their attention from the operations of the real masters of the State. In Communism, leadership comes from below, it is the leadership of the class and of the class-conscious party within that class. Individuals are important, but only in so far as they crystallise in definite actions the determination of the party and the class.

Communist leadership is objective in a full sense. It does not neglect psychology—it would be poor leadership if it did—but its psychology is an integral part of the appreciation of the concrete situation as a whole. The Communist is urged to think, not to trust. The ideal Communist is one who will know, even if he is isolated from all others as Dimitrov was, what has to be done, and does it. The ideal Fascist is one who will obey any order without question.

It is only necessary to compare a speech of Stalin's with one of Hitler's to see what a vast gulf divides the two conceptions of leadership. It is intended that the mistakes of Communist leadership, and there have been many, should be cured by deeper analysis of the situation, by better organisation, by the training of really class-conscious workers—not by using "psychology" to increase the self-confidence of the leaders and whip up the blind devotion of their followers.

* * * *

Enough has been said to show how far Mr. Osborn's book wanders from the path of marxism. Yet he is probably less to blame than those marxists who have never discussed the relations of Freud and Marx at all. Freudian influence is an objective fact, and is spreading slowly out from the bourgeois circles where it originated. Politically, it is a profoundly dangerous influence, paralysing action and tending to Fascism. Yet little or nothing is being done to combat it in this country.

The workers demand and have the right to demand a knowledge of psychology. If all they get is Freudian psychology, this is because English marxist writers have not applied themselves to the subject or even translated what has been written elsewhere. The one good effect the book should have—to provoke active discussion—may, it is hoped, produce serious analysis and criticism of the issues involved.

Footnotes

1. *Freud and Marx.* By R. Osborn. Gollancz, 7s. 6d. (Left Book Club, 2s. 6d.)

www.ingramcontent.com/pod-product-compliance
Lightning Source LLC
Chambersburg PA
CBHW070211290526
45789CB00002B/965

* 9 7 8 1 4 6 3 6 9 3 4 1 1 *